ADVANCE PRAISE FOR *Botany of Empire*

"Dwelling on the expansive field o[illegible] nes
ecologies, biodiversity, and extra-l[illegible] pire
and open to decolonial revision. *B*[illegible] hat
draws on and entwines feminist s[illegible] es, disability studies, and studies of colonialism in order to imagine both plant life and our collective livingness anew."

KATHERINE MCKITTRICK, Professor and Canada Research Chair in Black Studies, Queen's University

"Written by a leading scholar in the field of anticolonial feminist STS, *Botany of Empire* offers a beautiful guide that teaches the layers of anticolonial feminist approaches to science through the specificities of plants and botany. Combining a breadth of ambition with accessible, gorgeous writing, this book invites us to imagine a more just science in thought and feeling."

MICHELLE MURPHY, author of *The Economization of Life*

"Banu Subramaniam leads the reader on a lively trek across a historical landscape of colonial botany, narrating the 'great men of science' as not only relentlessly curious but conditioned by violent expansionism and misogyny. *Botany of Empire* is a compelling read for this Indigenous science studies scholar who stands on Linnaeus's grave every time I am in Uppsala. My aim, like Subramaniam's, is not to 'cancel' a revered European scientist who regretfully broke relations into taxa but to stand against colonial ideas of universal knowledge and progressivism co-constituted with the elimination of diverse peoples, knowledges, and moral orders."

KIM TALLBEAR, Professor, Faculty of Native Studies, University of Alberta

"*Botany of Empire* is an absorbing account of the many ways in which race, class, gender, and colonialism have shaped, and continue to influence, the plant sciences. This book will change the way you think about plants."

AMITAV GHOSH, author of *The Nutmeg's Curse: Parables for a Planet in Crisis*

BOTANY OF EMPIRE

FEMINIST TECHNOSCIENCES

Rebecca Herzig and Banu Subramaniam | Series Editors

BANU SUBRAMANIAM

Botany *of* Empire

PLANT WORLDS AND THE SCIENTIFIC LEGACIES OF COLONIALISM

University of Washington Press | Seattle

Botany of Empire was made possible in part by a grant
from the Samuel and Althea Stroum Endowed Book Fund.

Design by Mindy Basinger Hill | Composed in Fanwood

UNIVERSITY OF WASHINGTON PRESS | *uwapress.uw.edu*

LIBRARY OF CONGRESS CATALOGING-IN-PUBLICATION DATA

Names: Subramaniam, Banu, 1966– author.

Title: Botany of empire : plant worlds and the scientific legacies of colonialism /
Banu Subramaniam.

Description: Seattle : University of Washington Press, [2024] |
Series: Feminist technosciences | Includes bibliographical references and index.

Identifiers: LCCN 2023055153 (print) | LCCN 2023055154 (ebook) |
ISBN 9780295752457 (hardcover) | ISBN 9780295752464 (paperback) |
ISBN 9780295752471 (ebook)

Subjects: LCSH: Feminism. | Botany. | Science—Social aspects.

Classification: LCC HQ1155 .S874 2024 (print) | LCC HQ1155 (ebook) |
DDC 305.42—dc23/eng/20240116

LC record available at https://lccn.loc.gov/2023055153

LC ebook record available at https://lccn.loc.gov/2023055154

∞ This paper meets the requirements of ANSI/NISO Z39.48-1992
(Permanence of Paper).

Come, One and All, to Marvel and Contemplate
The Monumental Misrememberings
Of Colonial Exploits Yon.
KARA WALKER | *Fons Americanus*

Why can we imagine the ending of the world,
yet not the ending of colonialism?
KLEE BENALLY | *Rethinking the Apocalypse*

To love. To be loved. To never forget your own insignificance.
To never get used to the unspeakable violence and the
vulgar disparity of life around you. To seek joy in the saddest places.
To pursue beauty to its lair. To never simplify what is complicated
or complicate what is simple. To respect strength, never power.
Above all, to watch. To try and understand.
To never look away. And never, never to forget.
ARUNDHATI ROY | *The Cost of Living*

CONTENTS

Prologue Telling History *ix*

Introduction Through Linnaean Labyrinths: A Botanical Colonization *1*

PART ONE ROOTINGS

Chapter One The Botanical Sublime: Affective Ecologies and Plant Life *25*

Chapter Two The Coloniality of Botany: Reckonings with the History of Science *49*

INTERLUDES Fables for the Mis-Anthropocene—Chirp, Play, Love *60*

Revisiting the "Women in Science" Question: Diversity, Gender, and the Coloniality of Science *65*

PART TWO KINSHIP DREAMS | CLASSIFYING PLANT SYSTEMATICS

Chapter Three Categorical Impurative: Names, Norms, Normings *77*

Chapter Four Perhaps the World Ends Here: Spicy Embranglements in the Postcolony *103*

INTERLUDES Fables for the Mis-Anthropocene—Making a Little Trouble Everywhere *124*

An Ordinary Botany: Haunted Archives of Livingness *127*

PART THREE FLORAL DREAMS | SEXING REPRODUCTIVE BIOLOGY

Chapter Five The Orchid's Wet Dreams: Sex Told, Untold, Retold *133*

Chapter Six In the Dark Shadows of the Tree of Life: Sex, Race, and Reproduction *153*

INTERLUDES Fables for the Mis-Anthropocene—
The Queer Vegennials 170
International Council for Queer Planetarity:
The Botanical Debates 174

PART FOUR PANGAEAN DREAMS | MAPPING BIOGEOGRAPHY
Chapter Seven Botanical Amnesia: Colonial Hauntings in Plant Biogeography 183
Chapter Eight Like a Tumbleweed in Eden: Diasporic Lives of Empire 199

INTERLUDES Fables for the Mis-Anthropocene—
Love the Dandelion 213
A Cosmopolitan Botany: Tagore's Vision for Santiniketan 218

PART FIVE UPROOTINGS
Chapter Nine Vegetal Sublimations: Cartographies for Adisciplinary Sciences 227
Chapter Ten Dreams of a Lively Planet 236

INTERLUDES Fables for the Mis-Anthropocene—
The Memory Gardens 239
Against Eden 243
Abolitionist Futures: A Manifesto for Scientists 247

Acknowledgments 253
Notes 257
Index 303

PROLOGUE

Telling History

PLANET EARTH IS ON THE BRINK OF AN APOCALYPSE. AGAIN.[1] It faces massive extinctions, unbreathable air, undrinkable water, mountains of waste, chasmic inequality, and some very rich people plotting to escape the planet. Some call these times the *Anthropocene*. But not all *anthropos*, or humans, brought the planet here. Other histories lurk. Worlds of plants and animals are entangled in these histories. How to tell these stories?

How to tell the story of plant life on earth? The story of *Pangaea*, a conjoined single landmass about 250 million years ago. Tectonic forces pulled it apart. Plants and animals dispersed. Portions of the landmass evolved into new continents with astonishing adaptations, creating new kith and kin. European colonialism disrupted these differences and ruptured the planet anew.[2] World ecosystems collided in violence and turbulence, a veritable biological and cultural bedlam.[3] How to tell the fractured story of life, its histories in slavery and conquest, of lovers and friends, of revelations and revolutions?

How to tell a broken story? A story of a people that emerged out of Africa and dispersed into the rest of the world. Then ideologies invented new categories to carve humanity into biologically inferior and superior beings—master and slave, colonizer and colonized. In the ravages of colonialism, some looted at will the land and labor of others, only to retreat and beat the thunderous drums of nationalism to keep their ill-gotten spoils of conquest.

How to tell a mutilated story? Creatures lovingly named, nurtured into communities of multispecies life, only to be renamed, to be wrested apart, caged, maimed, killed, dried, displayed. Promises made, then broken. How to tell stories of betrayal? Playful critters in love, living in wild abandon, cataloged into myopic, restrictive categories. How to tell a story of limitations, of dependence and deception?

How to tell a mutant story? A primeval planetary soup billions of years ago ushered in life on earth, a dazzling array of exuberant biological life. How to tell the stories of lively mutations and frisky recombinations; a shared genetic alphabet spelling an astonishing diversity of life and living? How to tell the stories of winners and losers, competition and conquest, and also cooperation and care, symbioses and mutualisms, love and intimacy, solidarity and community? A lively story of a vibrant, vital planet.

How to tell a shattered story? Not to repair the broken history into fullness but to develop an epistemology and aesthetics that celebrates the fragmentary, partial, and incomplete.[4] How do we celebrate imperfection to crip wholeness and fully value "that which is collapsed, crushed, or shattered"?[5] Across vast time-scapes and mobile place-scapes, how to tell a story of belonging, partial belonging, unbelonging, rebelonging, and resolute nonbelonging?

History is telling.[6] We need multiple, competing, unresolved, fragmentary, shattered, partial, and contradictory tales. Rather than a singular story history tells, if we open up the archives to ask different questions, other answers lurk. These are also telling. And tell them, we must.

BOTANY OF EMPIRE

INTRODUCTION

Through Linnaean Labyrinths

A Botanical Colonization

MY BOTANIZING LIFE BEGAN under strange circumstances when my third-grade art teacher taught us a technique to draw people who were bending down. To my young asexual mind, it was odd but kind of neat. First, turn the paper counterclockwise by 90° and draw the head and torso, then turn it back to its original orientation and attach legs, feet (with appropriate clothing of course), and dangling arms. It was simple and quick. The teacher insisted that there be at least three people bending in every drawing. This posed a quandary for my young mind. People don't go around bending for no reason. I had to find a purpose. At first, I drew people sweeping the floor or exercising, worlds of clean and fit people. But these were weird activities for people in some landscapes. Then I hit upon the idea of flowers. People could bend to examine flowers, observe their structure, appreciate their beauty, and enjoy their fragrance. It allowed me to fill the page with many kinds of brightly colored flowers with many botanizing bent humans! With time I added specimen bags and simple instruments like magnifying glasses or rulers for measurements. Little did I realize that my botanical artwork followed in a long history of the sciences steeped in histories of sexism, racism, and colonialism. While many people across the world observed, studied, drew, painted, and used their knowledge of plants, only a few were allowed the privilege of a professional life in botany. This is a book about some of these histories.

Like most in the biological sciences, I learned little history during my training. Delving into botanical histories, I am amazed, outraged. Botany's foundational theories and practices were shaped, built, and fortified during and in the aid of colonial rule and its extractive ambitions.[1] Colonists were inevitably invested in the ambitions of empire, developing methodologies along the way. As I hope to show in this book, plant biology poorly captures the richness of plant worlds. We need alternative, richer epistemologies. This book is written from *within* the field of botany, and for all who share an abiding love of plant worlds and a thirst for justice.

Linnaean Labyrinths

Plants have long been important as medicines, herbs, and of course food. Royal gardens across the globe celebrated the vibrancy of plant worlds. But it was the advent of European colonialism starting in the fifteenth century that ushered in a more systematic and systematized knowledge of plants. Explorers roamed the colonies discovering a plethora of "new" and interesting plants.

In most histories of botany, one figure looms large: Carl Linnaeus (1707–1778), a Swedish botanist and taxonomist, and a colonial figure himself. He introduced a novel system of classification and nomenclature—a "sexual system" organized as a binomial with a genus and species name (for example, *Homo sapiens* for humans). He organized plants and flowers around an anthropomorphic imagery and in sexual binaries—male and female. In flowers, stamens became male and husbands, and pistils became female and wives; fertilization was likened to husbands and wives on their nuptial flower bed consummating a sexual union and marriage.[2] As Sam George argues, while earlier works upheld notions of female propriety, Linnaeus's *nuptaiae plantarum* (or the marriage of plants) opened up a polyandrous and polygynous sexual imagination where multiple husbands and wives were housed in flowers. This caused outrage, especially in a period where "the order of society was assumed to rest on the order of nature."[3]

By the eighteenth century, European women, usually elite gentlewomen, were cultivated into the feminized discourse of botany. Feminist histories document that women used the quotidian spaces of domestic gardens and fields to embrace the botanical and subvert feminine expectations.[4] Many women drew plants and painted them in their natural surroundings, and some even thrived as botanical artists. Botany and botanical art were exciting worlds. Botany was in the forefront of debates on female education, and writings in the eighteenth century reveal an "ambivalence in the process of the feminization of botany."[5] Ann Shteir documents powerfully that as botany marched toward becoming "modernized" and "scientific," the field embraced strategies to defeminize botany. She writes, "through textual practices and other means, women and gender-tagged activities were placed into a botanical separate sphere, set apart from the mainstream of the budding science."[6] By the mid-nineteenth century, the profession of botany was thoroughly a masculine enterprise and the ascendant male botanist its celebrated prototype. Likewise, we see the erasure of artisanal and working-class botanists.[7] As in other fields, women, once present in large numbers, were systematically excluded as the field emerged as a "science" and a male enclave.

One of the key insights of feminist work on the sciences is that even though nature is consistently gendered feminine (for example, "mother nature"), biology has persistently shaped the workings of nature as masculine and patriarchal—nature red in tooth and claw. The rise of botany transposed colonial views onto nature. No surprise, then, that there is more scientific work on competition than on cooperation, more on conflict than on coexistence, more on battle between the sexes than on joyful cooperative living.[8] Colonial worldviews ground branches of biology—both botany and zoology.

Botany flourished during colonial expansion as explorers "discovered" a treasure trove of plants during their global voyages. At its peak, botany was big business, fueling commerce and propelling the growth of merchant capitalism.[9] At the start of the eighteenth century, Australia and Antarctica were largely unknown to a Eurocentric world, and when colonial explorers in Africa, Asia, and the Americas described species they encountered, the diversity of those species astonished and overwhelmed. When Linnaeus began his career, "natural history was a mess, and people needed guidelines."[10] Drawing on the Greek myth where Ariadne fell in love with Theseus and gave him a ball of string to help him find his way out of the Minotaur's labyrinth, Jean Jacques Rousseau, an ardent botanist, praised Linnaeus's work as Ariadne's thread, allowing botany to find its way out of a dark labyrinth of colonial excess.

While the Linnaean system might seem simple in its binomial formulation, it was anything but. Its imagination and structures were fueled by powerful ideas about colonialism, race, gender, sexuality, and nation. The lasting legacy of this history is that all modern scientists are de facto Linnaeans. Plant names in botany today go back no further than his *Species Plantarum*, published in 1753, and all animal names in zoology begin with the tenth edition of his *Systema Naturae*, published in 1758.[11] Linnaeus's thread that showed the way out of the labyrinth of colonial botany continues to tether modern botany to colonial ideologies and sciences. Contemporary plant worlds, their names, and theories of histories, geographies, ecologies, and evolutions remain bound to the powerful hand of Linnaeus.

I use the labyrinth as a metaphor for the history of botany because it is both powerful and evocative. Martha Beck argues that the labyrinth is an ancient custom that isn't about solving a puzzle, but rather is a practice of mindful and meditative discovery through winding and curving lines.[12] Linnaeus attempted to resolve the labyrinth of biological diversity by organizing it into a simple system of nomenclature and classification. But in this system, the complexity of biological life and the richness of its worlds, especially the indigenous cultural contexts, were lost. Linnaeus built a thread that rendered biological life as a model of human gender, race, and sexuality as *he* saw it. In this book I follow the Linnaean thread back into the labyrinth. In retracing Linnaeus's steps, we come to understand the world he conjured up and appreciate what was lost and gained. We can meditate on botany's history, understand foundational theories in botany and the emergence of a botanical canon. We get to ask, Why this canon? Why is this the center of the narrative of the plant world? Importantly, how might we narrate otherwise? In challenging Linnaean sexual binaries, we challenge all binaries. Surely there are always more than two sides to every issue? Not a singular or binary view but a polyphonic, polybotanical imagination. In revisiting the labyrinth of infinite plant life, I urge us to see botany not as a site of the dark unknown of colonial scripts but as a site of joyful and playful exploration for flourishing botanical futures.

A Botanical Colonization

Years ago, I might have agreed that plants are an odd focus to revisit histories of colonization, but research for this book has astonished me. Understanding plant worlds through history reveals how central plants were to colonialism and vice versa. Yet this is not a comprehensive history of the colonial impact on the plant world. Rather, it is a retelling of botany through the histories of colonialism. It is a fascinating story about colonialism in all its varied avatars—ongoing settler colonialism, indigenous, postcolonial, and decolonial thought. I bring these in conversation with one another through plant worlds. Colonialism is an ideological, imperial, economic, and cultural project. The history of colonial botany is a story about more than plant worlds—how plants, animals, and colonized humans were used by and for the colonial project. By centering the plant, we see how colonists remade plants in their image, for their needs, consumption, and profit and for empire. While my focus is botany, revealing and resisting the hauntings of colonialism in botany reveals these same fissures in science as a whole.

Decentering the human is not a move to recenter the plant. I do not want to replace androcentrism with phytocentrism. Plants are anything but static; they are dynamic and evolving. In this era of a climate crisis, change is vertiginous. If colonization still informs our scientific knowledge practices, how might we undo these histories? We need rich epistemological and methodological landscapes to ground a countercolonial view of biology. We need to interrogate and challenge linguistic traditions that ground our theories, epistemologies, methodologies, and methods that shape botanical practices. Indeed, the clear boundaries between classificatory schemes of life on earth that shape biology classrooms—animals, plants, fungi, bacteria, viruses, and so on—are more porous than we acknowledge. Likewise, the idea of singular organisms and ecologies has given way to more complex understandings of assemblages, aggregates, microbiomes, ecosystems, networks, symbionts, holobionts, and so on.[13] I want to create bodies and landscapes without centers and peripheries and without hierarchical ordering.

The wise words of Audre Lorde are a central refrain throughout this book: "It is not our differences which separate women, but our reluctance to recognize those differences and to deal effectively with the distortions which have resulted from the ignoring and misnaming of those differences."[14] I expand this wisdom to understand that we do not need to collapse the diversity of life on earth into a quest for neatness, sameness, parity, or equity. As Lorde reminds us, we must celebrate difference by attending to our shared histories.

Key Conceptual Terrains

European sciences have transformed the majestic, deep history of plant time into the reductionist linear time of botany.[15] It is this transformation of plant worlds into the knowledge system of botany that interests me here. Today, plant worlds are botany. Botany is a powerful site that shores up one idea of nature.[16] It creates sites of purity such as "wilderness," and botanical technologies to help "tame" nature.

Some suggest that western science is itself best understood as an "ethnoscience"[17] and that appreciating its roots, routes, and evolutions are important and useful. Our knowledge production has been far too mediated by the politics of the academy.[18] The field of botany, like other fields, has "disciplined" itself into a narrow, myopic field, with a prescribed object of study (the plant world) and prescribed methods (the scientific method). Disciplinary education enables exploring the world from particular perspectives, reproduced generationally—perspectives that are taught, learned, rehearsed, practiced, remembered, and then replicated endlessly. As a result, I have much to unlearn as a biologist. Feminist science and technology studies (STS) reminds us that there are no sites of purity in the world, no sites exempt from the hauntings of colonial domination. How do we reckon with our colonial histories? Several key concepts that run through the book help weave the histories of colonialism through the natural and social sciences, the humanities, and the arts.

This book's foundation rests on refusing the binaries of nature and culture, instead embracing Donna Haraway's succinct and interdisciplinary term *naturecultures*.[19] Woven through the book you will encounter interdisciplinary vocabularies, theoretical approaches, and analyses, as well as multiple genres of writing and varied tones. We must experiment with alternate genres of writing and value fragmentary insights, momentary glimpses, partial views, imperfect biologies, and transient ecologies as important grounds for understanding and theorizing. If the coloniality of science shapes the form of a scientific paper, a book is usually squarely in the humanities. In writing a book about botany as a naturecultural field and drawing on and integrating authors and scholarship across academe and outside it, I take an epistemologically radical stance. I offer a multitude of genres—from disciplinary forms of articles and essays, to autobiographical and biographical entries, memoir, manifesto, fables, fiction, and speculative fabulations.

Interdisciplinarity necessitates thinking critically and questioning one's assumptions. For example, as a biologist confronted with the idea of native and foreign plants, I use my critical thinking skills to interrogate definitions of *native* and *foreign*. Are these historical terms? As we will see in the later discussion of invasion biology, historicizing botany allows us to recognize these as imprecise, indeed political, categories rather than natural or biological ones. Is the natural world organized into species? No; these are human constructs. To be sure, such conceptions can be immensely helpful, but they are also deeply constraining and sometimes misleading. Histories and contexts matter. In following Linnaean threads into the labyrinth of botany, I attempt to understand how and why biological concepts came into being. I hope this book demonstrates the immense power of an interdisciplinary education and why such approaches produce more robust knowledge about the world.

This book focuses on reckoning with the histories of colonialism. While I explore these histories in greater detail in chapter 2, some conceptual tools are critical. Colonialism isn't an event or a historical blip of actions but an enduring installation.[20] As Edouard Glissant succinctly observes, "the West is not a place, it is a project."[21] Understanding colonialism as a project allows us to see its vast infrastructures in academic disciplines. It is thus useful to talk about *coloniality*, the embedded histories of colonialisms that persist.[22] Infrastructures of coloniality include not only the epistemologies, methodologies, and methods that structure disciplines but also infrastructures of sex, gender, race, and sexuality.[23] Importantly, coloniality's infrastructure, grounded on colonial ideas of race and gender, erased other models of social organization and myriad local systems of knowledge the world over. Robin Wall Kimmerer frames indigenous ecologies as maintaining good relations in everyday life.[24] She points to an emerging consensus about indigenous knowledge systems as fundamental to conserving biodiversity.[25] Globally, indigenous peoples inhabit and maintain areas with some of the highest biodiversity on the planet and are engaged as partners in many biodiversity conservation measures.[26] While Traditional Ecological Knowledge (TEK) is recognized as having an equal status with scientific knowledge and being "an intellectual twin to science,"[27] it is consistently marginalized by the scientific community.[28]

In working on this project, I came to adopt the term *embranglements* (the state of being embroiled or mired in something). I find this older term more useful for discussions of colonialism. While terms like *entangled*, *intertwined*, and *implicated* imply the interconnections of worlds, *embranglements* also implies tension within—shaking, wavering, confusing. Interdisciplinary embranglements are always fraught, capturing the difficulties of interdisciplinarity. We need to work through our embrangled histories.

In approaching interdisciplinarity, I draw from literature because it best captures history's horrors. For example, echoing Avery Gordon's *Ghostly Matters* on haunting in sociological life, the figure of the ghost and its hauntings, a theme that was central to my earlier book *Ghost Stories for Darwin*, haunts this book as well. But I want to do more than recognize and listen to ghosts: I want to retool botany with concepts that can deal with our haunted colonial histories. So much of botanical history remains grounded in internal histories of the west and the biosciences. Lost, forgotten, and erased are the genealogies of women of color feminists, indigenous feminists, and postcolonial, diasporic, crip, queer, and trans feminists, who have always written more syncretic symbiotic stories that do not privilege the "human." In bringing feminism and botany together, I trace how botany's colonial roots shape its foundational language, terminology, and theories; the field remains grounded in the violence of its colonial pasts. Collaborations between feminist, indigenous, and biological thought can help us work toward more just planetary futures. Recent work by biologists such as Cleo Wölfle Hazard, Jessica Hernandez, Robin Wall Kimmerer, Meg Lowman, Joan Roughgarden, and Kriti Sharma, to name a few, remind us of how critically intertwined the personal, scientific, and political are to a life in biology.[29]

In reconstructing history—of botany, feminism, and the planet—I draw on a central concept from Toni Morrison in *Beloved*, "rememory," a term she uses as both a verb and a noun, that which "turns the present of narrative enunciation into the haunting memorial of what has been excluded, excised, evicted."[30] "Rememory," as Viviane Saleh-Hanna argues, "is preserved in institutions, branded upon their violently structured bureaucracies and practiced upon the bodies of the colonized by the bodies of colonizers: a specter is haunting modernity—the specter of colonialism."[31] In her "Black Feminist Hauntology," she writes that "the term 'ghost' neither confirms nor denies the metaphysical. It simply invokes a framework in which terror and unpredictability, grief and unrest, guilt and injustice, ancestors and demons can be called upon to empower and liberate us, not from the fact that we have been violated or even that violation continues, but from a condition of inability to locate the heart and soul of the problem." Jeong-eun Rhee reminds us that in moving across past, present, and future, rememory connects time, space, matter, and histories. As she evocatively argues, rememory isn't just about theory but encompasses affective experiences where the breath of the wind, the fluttering of the wings of dragonflies, or the stirring of leaves can become a haunting and powerful presence.[32] For Morrison, the past does not remain in the past but emerges as a site where we can make deeper discoveries. In a language "indisputably black," Morrison opens up spaces for those historically excluded. To Morrison, ghosts do not return; they are "immanent to space."[33]

While Morrison's original concept was largely about human worlds, I'm expanding her work here to the nonhuman and to the realms of the genetic and ecological, as well as the vast generational weight of plant life and adaptation. What is powerful about the concept of rememory is that it opens up the past, especially the lessons we have forgotten, unlearned, or never been taught. It echoes what Christina Sharpe calls "wake work," a way of reflecting and of "re/seeing, re/inhabiting, and re/imagining the world."[34] Opening up registers of memory, rememory forces us to contend with the histories of colonialism, racism, heterosexism, ableism, and misogyny and to ask how these histories have shaped the landscape of scientific theorizing. Rememory can help us recognize the profound botanical amnesia that produced xenophobic concepts such as invasive species, "discovery" of plants long known to natives, and translating the exuberance of plant reproduction into the decidedly human registers of "sex."

As we rememory the history of botany, the past opens up. Histories show how and why academic disciplines and subdisciplines, developed and consolidated through colonialism, have produced structures of coloniality—nomenclatures, taxonomies, epistemologies, methodologies, methods, ontologies, and theories sanctified by liberal logics. As this book chronicles, the original colonial bioinvasion is followed by a science of invasion biology. Linnaean "marriage of plants" produced modern reproductive biology and its battle of the sexes. Colonial bioprospecting laid the conditions for modern biopiracy. If scientific stories narrate the history of life out of Africa in the language of race, species, populations, or individuals, then rememory opens up our ability to explore the texture of those memories in the flesh, in the sinew, in the pores of the living and the dead—the ghostly afterlives of Malthus, Darwin, Humboldt, and Linnaeus and new tales of life on earth.

Similarly powerful is Sylvia Wynter's insistence that we move beyond the binaries of colonizer/colonized and perpetrator/victim because such oppositional models force a view from either the celebrant ("European") or the dissident (what Wynter calls "Native"), locking us in the same colonial order and colonial framings.[35] This is where rememorying the complex histories of naturecultures is immensely useful. It allows us to unlearn our disciplinary narratives about natures and cultures and instead commit ourselves to rememorying new genealogies of a naturecultural planet—through fracture and union, through conquest and liberation, through competition and cooperation—to produce a dizzying vista of thoroughly embrangled lives. For example, how did the tumbleweed, a foreign and indeed invasive plant, become an icon of the American West? Why are some plants reviled and others celebrated? Rememorying plant life through naturecultures helps us narrate embrangled lives under and in the wake of slavery, colonialism, conquest, and servitude, helping us imagine more just futures.

Tracing the colonial roots of botany opens up questions of decolonization. Rather than critique from without, I choose to work from within, to excavate botany's disciplinary formations and foundations and expand its limited and myopic sphere of "nature" into new articulations, theories, and concepts that can better account for our embrangled worlds. Rising beyond the tendencies to conceptualize groups as individual, population, species, genus, variety, class, phylum, or kingdom, rememory foregrounds networks of relationality that emerge from a hypermobile, cross-pollinated, interbreeding world.[36] For example, in a naturecultural world, plants are often assigned ethnonational groups even as they develop new ecologies in changing networks of botanical and political geographies. In the United States, for example, we identify some plants with such names as Japanese knotweed or Chinese privet and yet anoint the Georgia peach as American even though it is of Chinese origin. Repeatedly, desirable objects become US American while the undesirable retain their foreign monikers.[37] The majority of US crops are plants of foreign origin, while most insects that cause damage are considered native.[38] We need to historicize botany and our accounts of plant life in their complex global ecologies of relationality if we are to have any hope of scientific explorations that do not merely reinscribe histories of colonial investments. In short, we need to queer botany.

Queer Studies and Disability Studies

In the course of my work, the fields of queer studies and disability studies emerged as important interlocutors. Both challenge binaries: abled/disabled and straight/queer. In challenging the binary classification of bodies as abnormal or deviant,[39] they invite us into rich landscapes and worlds with variety and diversity rather than pathology. The field of disability studies chronicles how science and medicine were critical to transforming ideas of biological variation, understood within realms of the moral, spiritual, and metaphysical during the nineteenth and twentieth centuries, into medicalized bodies. Under "the medical model," disabled and queer bodies were pathologized as lesser, deviant, and undesirable, with profound consequences. Eugenic laws, for example, were instrumentalized across the world to sterilize, institutionalize, and at times even eliminate queer and disabled bodies. The history of eugenics is a grim reminder of the power of science, medicine, and the state, especially when all align.[40]

Medical sciences came to anoint themselves as saviors who could help individuals *overcome* or who could *cure*.[41] It is impossible to understand this framing of disability without recognizing that racial capitalism has narrowly shaped our understandings of what counts as meaningful work and productivity.[42] Feminist economists powerfully demonstrate how caring labor has been long been relegated to unpaid work and the private domestic sphere—even inside academia.[43] Histories of care work remain deeply feminized and racialized. Indigenous, disability, and queer rights activists remind us that caring for each other and the planet is critical for life and for social and planetary justice.[44]

Four concepts in particular—natural, normal, unnatural, and abnormal—form a powerful matrix of inclusion and exclusion.[45] The link between binaries of natural/unnatural and normal/abnormal are resonant frames throughout this book. The solution is always about finding ways to "help" and to restore ability of some kind, thus reinforcing the normal and the normative as desirable spaces that all must emulate. But who sets the standards? For example, mobility is an issue only when modern infrastructures insist on narrow or heavy doors, inaccessible staircases, or spaces that make it impossible for some to navigate.[46] Hearing and seeing worlds also dominate our lives. In contrast, accessible practices and thoughtful infrastructure open up the world for all. As activists powerfully demonstrate, the problem is not the excluded *individuals* but the built infrastructures that exclude.[47] And when anthropocentric concepts are transferred into plant worlds, botany also becomes a site that reinforces the normal and natural.

Crip theory eloquently captures ableism with the term *supercrip*. As Eli Clare writes, the supercrip is one of the dominant images of disabled people. We are taught to celebrate the boy without hands who bats well, or a blind man who hikes the Appalachian Trail, or an adolescent girl with Down's syndrome who learns to drive. The nondisabled world is suffused with such stories where resilience against all odds is celebrated—a visible and repeated lesson that disabled people must overcome disability to be celebrated.[48] Disability is an important topic within botanical worlds because the plant literature repeatedly notes, often with alarm, the immobile, stationary, and rooted nature of plants. And yet plants manage perfectly well in living, transporting their pollen and seeds. Their indeterminate growth means their branches can fill the canopy, and their roots grow deep and wide. Mobility is a mindset of the able-bodied human as prototype, and in built worlds that restrict rather than include. This includes scholarly and political exclusions of the disabled communities from environmentalism as well as physical exclusion from gardens and national parks.[49]

Human life spans dominate anthropocentric views of the world; plant lives, in contrast, can be considerably shorter or longer. The most violent and misarticulated impact of colonization is what Sumana Roy refers to as the "substitution of forest-time by this imported industrial idea of time."[50] The term *crip time* from disability studies captures how disabled, chronically ill, or neurodivergent people experience time (and space) very differently than able-bodied/minded people.[51] There is a difference between crip time and "normate" time.[52] Crip time captures disabled peoples' different experiences of time in the world. These ideas link disability studies with critical animal studies and critical plant studies.[53] Disability studies has taught me to cringe when plant super-cripness is repeatedly invoked in recent literature on the celebration of plants—*They cannot move, and yet they can do so much!* The language of movement and ableism is striking in the plant literature, especially in the recent turn to plant and tree love.

Queer theory and the field of queer studies also shape this book. Challenging heterosexuality and reproductive (hetero)normativity, queer studies emphasizes the necessity of thinking about sexuality not in terms of bodies or identity but as a field of power.[54] The term *queer* has grown capacious with time. Eve Sedgwick, one of the founders of the field, defines *queer* as "the open mesh of possibilities, gaps, overlaps, dissonances and resonances, lapses and excesses of meaning when the constituent elements of anyone's gender, of anyone's sexuality aren't made (or can't be made) to signify monolithically."[55] *Queer* as a verb is also central as method: to make strange and to question what we know. To think, read, or act queerly is to think across boundaries, beyond the normal and the normative; to explore the spaces deemed marginal, vulnerable, precarious, and perverse;[56] to celebrate, in Angela Willey's words, "queer feminist desires for new modes of conceptualization and new forms of belonging."[57]

Like crip time, queer time captures how queer people have had to contend with a world where heterosexual (and cis-bodied) expectations of marriage, children, and family were, and are, closed to many. Transgressive moments of sexuality, such as coming out for queer people and transitioning for trans people, warps time and prevents the unfolding of linear time.[58]

Both queer time and crip time remind us of how expectations of the normal link to experiences of time and space, and why challenging normative ideas in describing plant worlds is productive.[59] After all, plants are forever forced into human time for science and commerce—botany, agriculture, horticulture, and plant biotechnologies. As plant lovers and passionate interlocutors with plant worlds, we must reckon with this history.

Both queer and disability studies have blossomed into ecological thought. Queer and trans ecologies have pushed for a more expansive understanding of the world in terms of rethinking ethics and multispecies entanglements. How do we live *with* and *in* the natural world without exploiting it? Rather than focus on the natural or seek a nostalgic return to the past, queer and trans ecologies help dismantle exclusionary structures of western science. Rather than fixate on an "ideal" or "right" nature, queer and trans ecologies stress multiplicity and opening up space for genderqueer and nonconformist bodies in many senses of the word (human, animal, plant, land, water).[60] Similarly, links between disabled ecologies and environmental devastation allow us to see how key concepts from disability studies—loss and limitation, vulnerability, interdependence, and adaptation—might offer key lessons for accessible futures for myriad disabled beings and impaired landscapes.[61] Queer, trans, and disability studies thus offer us rich frameworks to imagine botanical futures.

In reconstructing histories—of botany, feminism, and the planet—I am exploring new genealogies that recognize colonialism as a specific, and not inevitable, historical intervention whose legacies are ongoing. Yet as Klee Benally asks in the book's epigraph, if the desire to colonize was first born in the imagination, why can we not imagine its end? So much of STS remains grounded in the west and the biosciences. In disrupting this story by bringing feminism and botany together, we see how botany remains grounded in the violence of its colonial pasts. Collaborations between feminist, indigenous, and biological thought can help us work toward more just planetary futures.

Outline of the Book

This book is written for multiple audiences. It brings together the natural and social sciences and the humanities and arts to showcase how interdisciplinary approaches can transform our understanding of the "natural" world. In historicizing biology, we confront the imperial legacies that shaped disciplinary silos, with their singular focus and myopic visions, and reckon with this history to imagine a more capacious biology.

My main goals are threefold: explore how botany was shaped by colonialism; demonstrate how that history endures in contemporary botany; and ask how we might undo these legacies to imagine an interdisciplinary and countercolonial botany that is less anthropocentric and more empirically attuned to plant worlds.

At its core, the book advocates for the critical need for work across academic disciplines. The sciences need humanistic inquiry, and the humanities need the sciences. The future of the planet depends on it. For biologists, this book historicizes the field, making a familiar world unfamiliar. For social scientists and humanists, it introduces botanical worlds in a new idiom, making unfamiliar worlds more familiar. An interdisciplinary approach is critical for the problems we face. The natural world and its myriad environmental crises cannot be adequately understood by the tools of botany alone. In opening up the worlds of botany and feminism through interdisciplinary approaches, we see new multispecies possibilities.

In reckoning with the histories of science, how might we decolonize botany? I start with the answer that we can never decolonize botany within disciplines and institutions that remain deeply colonial. The histories of settler colonialism, postcolonialism, and decolonial thought all offer important lessons. Ultimately, I approach decolonization much like feminism, as an engaged, reflexive praxis, an intentional movement toward more just futures.[62] Since much of colonization is a top-down process, decolonizing efforts must necessarily be a bottom-up process. Decolonization cannot be a singular project. The book is inspired by multiplicity, hybridity, interdisciplinarity—epistemologies and methodologies drawn from many disciplines, multiple methods to engage with the plant world, and multiple genres of writing. Decolonization is also necessarily dynamic. Powerful forces that benefit from colonial histories have undermined movements for justice and will continue to do so. For this reason, I draw heavily from scholars and activists who are attuned to methodological landscapes for questions of difference—not in the ostensibly objective register of disembodied and disengaged knowledge but in one that has social justice front and center.[63]

This book means to provoke an overdue conversation. Because the work is a historical and colonial reckoning, I have retained the term *botany*, but you can easily substitute newer terms like *plant sciences* or *plant biology*. I retain the older term fully appreciating that both the term and the formations of the disciplines of botany (and zoology) have been in decline.[64] Instead, we see the study of plants within new and broader interdisciplinary fields like general biology, integrative biology, organismal biology, ecology and evolutionary biology, and molecular and cell biology.[65] But whatever the name, the same histories and issues persist.

Botany is a vast field—from the planetary, ecosystem, and organismal levels right down to the molecular. Within the "pure" sciences, botany has developed areas of specialization: plant anatomy, biogeography, biomechanics, cell biology, ecology, evolution, genetics, molecular biology, population genetics, physiology, reproduction, systematics, and taxonomy. These subfields have related but unique histories. A study of the whole field proved too much for one book, so I focus on just three subfields: plant taxonomy, plant reproductive biology, and plant biogeography. Plant taxonomy provided *order* that colonizers sought to organize the natural world. In systemizing the world into categories and an evolutionary tree for life on earth, plant taxonomy is a critical node of colonial botany and its enduring afterlives. Plant reproductive biology chronicles how the imaginaries of gender and race under colonial sexuality were imposed on plants. Reproduction, central to theories of Darwinian evolution, is the bedrock of modern biology. Finally, understanding plant biogeography through invasion biology centers questions of space and time. Do organisms *belong* in a particular place and time? What work do concepts such as native and foreign do? The questions are central to our embrangled histories. We travel the Linnaean labyrinth in five parts.

The first part, "Rootings," grounds the book in a broad framing of key challenges and delights of studying plant life and living. Chapter 1 tracks the study of botany by intertwining a brief history of the field with my personal reflections on coming to a life in biology as a postcolonial child. Both stories are grounded in the idea of the "botanical sublime." Chapter 2 is a theoretical chapter on history and colonialism. History is no innocent field; internalist histories of botany seldom acknowledge the histories of colonialism, slavery, or conquest. Drawing on recent work rethinking the field of history, I explore colonialism in its many avatars across the globe. I describe the varied analyses and stakes of settler colonialism as well as indigenous, postcolonial, and decolonial thought. This chapter serves as an introduction to histories and schools of thoughts I draw on throughout the book.

The next three parts focus on the three main case studies.

Part two, "Kinship Dreams," explores the fields of plant nomenclature, classification, taxonomy, and systematics, fields that organized and brought order to the plant world. This ambitious history spans from the early beginnings of botany all the way to the modern plant sciences. Through colonial exploits during the "Age of Exploration," colonists went in search of botanical resources. The vast infrastructure of botany that ensued was grounded in liberal logics—the (always unwelcomed) "exploration" of colonized lands leading to claims of "discoveries" well known to natives of those lands. Moving in broad strokes through a large expanse in time, I show how rooted modern biology remains to these early standards of plant naming. White colonists, some very brutal, continue to be celebrated in plant scientific names. Each species' "holotype specimen," selected by the original author when the species was named, described, and published, often still remains housed in a western herbarium. A recent study by scientists and curators from herbaria across forty countries paints a damning picture.[66] Of the 3,426 herbaria in the world that house approximately 400 million specimens, over 60 percent of the herbaria and 70 percent of the specimens are located in developed countries with colonial histories. Herbaria in the United States and Europe house twice the number of species that occur in these countries, a colonial appropriation of large amounts of plant diversity. In contrast, Africa and Asia herbaria house far fewer specimens than are collected there. Of the specimens with digital images, 80 percent are held by European and North American institutions, not all accessible. In a profound irony, the collections of colonial botany ensure that there is an inverse relationship between regions where plant biodiversity exists in nature and where it is housed in herbaria! Recent efforts of digitization and decolonization have done little to alleviate colonial legacies. Colonial-era practices endure.

I tell the story of the plant taxonomy through two different histories. Chapter 3 rehearses the history of plant nomenclature, classification, and taxonomy. Chapter 4 uses South Asia as a particular case to show how the afterlives of colonial botany shape modern nations. The importance of plants and their medicinal and therapeutical values—then and now—continue to shape the modern plant sciences. These legacies reveal the heavy and enduring role of botany's coloniality in their postcolonial and neocolonial legacies.

Part three, "Floral Dreams," explores the field of plant reproductive biology. If race and nation emerged front and center in the case of biogeography, sexuality emerges as a critical node in reproductive biology. I have, as I hope you will, come to recognize the profound androcentrism that grounds scientific views of plant reproduction.

The innovation of sexual reproduction is purportedly an innovation for producing *variation*, the selective terrain and playground for natural selection. Londa Schiebinger argues that the grounding of *sex* as a central attribute of plants is an accident of history. The scientific revolution and the revolution in sexuality and gender came together to elevate plant sexuality as a central focus of botany. Cultural and social ideas of sex and gender shaped scientific understandings of plant sex.[67] Carl Linnaeus, the "father" of plant taxonomy, connects the worlds of plant reproduction and nomenclature, where sexual organs (modeled around the human) come to shape the classification and organization of the plant kingdom. Linnaeus gave primacy to plant sexuality, and his "'scientization' of botany coincided with an ardent 'sexualization' of plants."[68]

I explore plant reproductive biology in two very different ways. In chapter 5, I explore how plant biology is narrated. How and why do plants have binary sex and gender? How did the conceptions of (western) humans come to shape plant sex, sexuality, and gender? In detailing why and how plants have sex, we must ask whether plants actually have sex. Is *sex*, modeled around human reproduction and its embrangled histories, the best term for what plants do? I argue that it is not and offer alternate models of theorizing plant reproduction. Chapter 6 explores the shared histories of sex, race, and reproduction across plant and human worlds through histories of colonialism. Sex and race are deeply intertwined in these histories, and their conceptual frameworks in the colonial mindset travel into plant worlds. Using the tree of life as a metaphor for the evolution of life on earth, I show how theories of difference have shaped colonialist ideas of reproductive and evolutionary biology of human and plant alike.

Part four, "Pangaean Dreams," explores the field of plant biogeography through the idea of invasion biology. This idea is predicated on a binary view of nature *in place* and *out of place*. Deeply racialized, the concept and subfield of invasion biology stoke xenophobic alarm of a world increasingly out of place. I frame the discussion of invasion biology through histories of colonialism as an act of botanical amnesia. I juxtapose anxieties about invasive species today with European colonialism that ushered a massive and grand reshuffling of global biota—indeed, the original bioinvasion! Alien species in the colonized worlds are, in fact, legacies of colonial botany. While widespread ecological devastation and species extinctions have occurred, scapegoating foreign species is poor history.

I explore plant invasion biology in two chapters. Chapter 7 deals with the troubled definitions of the native. In light of colonial botany it is impossible to malign the foreign. I trouble recent attempts to reinvent the native through contemporary politics—invasive species as colonizers, refusing invasive species as an act of decolonization, or invasive species as an enemy of local culture. Chapter 8 explores questions of invasion biology through the language of hybridity and diasporic life. In short, how should we understand the native? Is the native a product of migration and hybridity highlighted in diaspora and postcolonial studies or of the settler colonial and the native of indigenous studies? I refuse this binary choice that pits postcolonial and indigenous studies against one another. A decolonial botany must confront histories of land, of violence and conquest, but it must also reckon with the colonial violence that produce colonized peoples, migrants, and refugees. We must contend with the multiple histories of colonialism. We must overthrow a racialized and reductionist botany that celebrates native seeds without lands, peoples, cultures, and their histories.

Part five, "Uprootings," concludes the book with key lessons in the history of colonization and botany. If there are colonial logics, what are decolonizing logics? How do we undo colonial logics, however modest such an enterprise may be? Perhaps we can begin with a rejection of the academic story of "two cultures" where the humanities and sciences are separated.

I offer interludes at the end of each part. These are spaces of cultivation and contemplation, exploring alternate imaginations and projects of decolonization—fiction, fables, biography, thought pieces, and manifestos. They are meant to provoke, engage, trouble, and imagine the world anew.

Botanizing in the (After)Lives of Empire

A true biological reckoning acknowledges that we are a damaged planet, all refugees of a ravaged naturecultural colonial past, seeking to salvage our naturecultural present and futures. The constructions of natives, aliens, migrants, and refugees are all political constructions of the unequal afterlives of empire. The ravages of empire have transformed not only human and cultural landscapes but also ecological ones; no species is well adapted anymore. We are all displaced, no longer living in the worlds we grew up in, our environment no longer familiar; we are all refugees, albeit in very unequal and hierarchical worlds. The rise of the global Right bespeaks a global anxiety about place. But rather than focusing on nativism, thinking in and out of empire reminds us that we are all adapted to worlds that no longer exist at *home*. What feels like home could be thousands of miles away, on another continent. Reckoning with the false borders and boundaries of nations and nationalisms are not only about human worlds but also about our co-inhabitants of the planet—the plants that feed us, the fabrics that clothe us, and the lumber that often houses us. We need new naturecultural imaginations for our ruderal lives.

How do we undo the coloniality of power that ushered in a global genocide, ecocide, and epistemicide? We must think about rematriation and reparations.[69] Where do our herbaria specimens come from? If stolen, how do we return them? While we figure that out, how do we make it possible for communities to engage with their rightful inheritance through free funding and access?[70] When western scholars do fieldwork in formerly or settler colonized nations, what are the terms of engagement? Is permission and collaboration sought? Who should give permission? When permission is given, are institutions and infrastructures built? How is power shared? If permission is denied, do scientists accede? Such ethical and political considerations must ground scientific methods and methodologies. How do we treat students from colonized lands in botany? Do we teach colonial histories of botany? Shouldn't we educate all our students on indigenous botany, ethnobotany, queer botany, postcolonial botany? How do we empower students to imagine anew? Throughout the book, I examine concrete ways we can rethink the disciplines. A few examples:

- We must decenter a history of biology centered on the west.
- There is no universal template. Decolonizing is not a thing or prescription but an ongoing process with indigenous and formerly colonized communities alongside botanists, curators of herbaria, plant lovers, and scientists of horticulture, agriculture, and plant breeding. If botany started with a set of elites who imposed it on the rest of the world, decolonization cannot replicate these power relations. It must be made anew, collectively.
- Colonialism was not only a genocide but also an ecocide and epistemicide. Decolonizing necessitates many solutions, at many scales and geographies. There is no one solution for all.
- We need to recognize the rights and responsibilities of all peoples. We cannot continue to practice "parachute" science (where botanists pop into parts of the world, acquire biological samples, and pop back to the west for analysis and glory). We cannot presume power or consent. When we work with others, it must be through mutual collaboration. When groups say no, we must honor it. Similarly, we must welcome ideas from others even if they seem incommensurable with our own.[71]

- Colonial science refused to recognize local knowledges in the colonies as science. Locals had long cultivated crops, herbs, spices, and medicines. For example, the bark of the cinchona tree was an old and popular remedy in Peru. Yet the species was named by Linnaeus in 1742 after the Countess of Chinchon, who brought it back with her to Spain. Botany is filled with such tales. At best we reserve the term "ethnobotany" for local knowledges. Can we retheorize ethnobotany as science, or retheorize science as a form of ethnobotany?
- While colonialism destroyed indigenous ways of knowing, it also appropriated and incorporated local knowledges into its repertoire. In chapter 4 on the *Hortus Malabaricus* we see how local and subaltern knowledges were appropriated as science but their indigenous roots forgotten and erased. Rejecting botany wholesale can mean losing subaltern knowledges. We must support research on the colonial roots of botanical ideas, theories, concepts, and practices.
- Botany is a selective knowledge, and we need to recognize the strategic choices of colonists. For example, Dutch botanists (exclusively men) who learned about the abortifacient properties of plants from local women did not transmit that knowledge to their naturalist colleagues or women back in Europe.[72]
- Most academic disciplines, including botany, STS, and feminist studies, center the west and whiteness in their analysis. How do we nurture non-western sciences as science?[73]
- How do we resist co-optation through neoliberal appropriation or institutionalization of our efforts at epistemological and societal transformation?[74]
- Colonialism was not built in one day; decolonization will take longer. How do we develop a practice of strategic patience?
- Decolonizing is an enduring commitment, a historical reckoning. It requires a sustained and persistent commitment, against all odds.

When presenting this work, I repeatedly encountered the argument that while decolonization is important, it will "set us back." For example, take the case of renaming plants so that racist and genocidal colonists are not honored in plant names. Plant renaming is not alien to the field of botany; plant systematics routinely reclassify plants based on new evidence. Yet in the context of decolonization, renaming is considered anarchy! Proposals to rid botany of such names have faced steep resistance.[75] The urgency narrative of progress in normate sciences derails decolonizing efforts. I have to ask: Whose time? What is being lost except a celebration of racists and white supremacists? So what if we lose some time? Justice, like knowledge, is surely a worthy goal.

Much of plant writing describes plants in ableist terms, as rooted, immobile, and nonsentient. Yet, as crip theory challenges us, plants are very much alive and sentient. For example, Michael Pollan raises the provocative notion of grasses having colonized humans—look at all the time that humans spend on lawn care![76] Renewing attention to plants and plant biology offers us new vocabularies for life and living, inviting us to engage into naturecultural worlds with less androcentrism and greater humility.

In *Ghost Stories for Darwin*, I confronted the figure of the abject ghosts of scientific reason and racism. Rather than repel or silence the ghosts, I engaged with them to understand a long-repressed history. Having confronted these colonial, eugenic, racist, and misogynist histories, I can now see past the fading specters of Darwin, Linnaeus, Humboldt, and Malthus. I see other ghosts, enchanted ghosts—the lively ghosts of a vibrant and vital botanical past. The task before us is a renewed imagination, rememorying the many paths not taken, the many futures that were once possible. To be sure, there is no purity, no Eden to return to—yet we still have exuberant, enchanted, teeming landscapes of radical botany, queer botanical worlds teeming with anticipation and promise. As we travel through Linnaean labyrinths of historical botany, we can better recognize the fraught embranglements that bring us here. If biological models were forced upon plants so they would resemble colonial humans, could we discard and even reverse this view? What if we worked from the biologies of plants to reimagine plants, and from there to rethink the human?

PART ONE

Rootings

When is it time to dream of another country or to embrace other strangers as allies or to make an opening, an overture, where there is none? When is it clear that the old life is over, a new one has begun, and there is no looking back? From the holding cell, was it possible to see beyond the end of the world and to imagine living and breathing again?

SAIDIYA HARTMAN | *Lose Your Mother*

CHAPTER ONE

The Botanical Sublime

Affective Ecologies and Plant Life

> Life, too, is like that. You live it forward but understand it backward.
>
> ABRAHAM VERGHESE | *Cutting for Stone*

> I once thought I knew what nature writing was: the pretty, sublime stuff minus the parking lot. The mountain majesty and the soaring eagle and the ancient forest without the human footprint, the humans themselves, the mess. . . . The problem is, the Nature/Human split is not a split. It is a dualism. It is false. I propose messing it up. I propose queering Nature.
>
> ALEX CARR JOHNSON | *How to Queer Ecology: One Goose at a Time*

HOW DID I DEVELOP A LOVE OF BOTANY? I grew up an urban child in India. I grew up in a jungle—a concrete jungle. It was magical, a cornucopia of sensoria. A world filled with heavenly smells, sensual textures, delightful play, eclectic sounds, vibrant sights. The sweet nightingale competed with the blaring car horn, the scent of the jasmine with the pungent disinfectant, the spicy potato *bajji* with the sweetest mango, the gorgeous *gulmohar* aflame in red with the vibrant saris on the street. Botany was everywhere, the luscious memories of a joyful and adventurous childhood.

In recent years, I have been consumed, once again, by the world of plants. My current fascination for plants is not as a botanist using the tools of botany

I was taught in my biological education. I am also not abandoning a rigorous botanical education and its tools in the name of some naive, sentimental plant love. Rather, I am trying to bring to bear the vast insights and tools of feminist, postcolonial, and indigenous science and technology studies (STS) to revitalize my fascination for the natural world, and for plants in particular. At the same time, I am trying to enrich my life in the humanities by embracing the vibrancy and vitality of plant worlds. In this new work, I am attempting to reestablish my love of plants through a cartography for decolonizing the field of botany. In preparation, I have immersed myself in the worlds of colonial plant biology—their histories and historiographies. My graduate biological education in the United States promised me a world of botany as objective, dispassionate, and value-free knowledge. I was taught that good scientists abandoned their emotions outside the doors of the laboratory or the field site. Growing up in postcolonial India, I came to believe that science, the scientific method, and the scientific revolution primarily emerged in the west. In independent India, science was the vehicle by which to enter the hallways of modernity and a brighter future. Neither of these core experiences prepared me for the histories I discovered. Reading anticolonial, postcolonial, and indigenous histories of plants, I was outraged. The process has been thoroughly unexpected and disquieting, bringing up strong feelings of utter surprise, shock, awe, wonder, anger, disgust, contempt, joy, fear, pride, bewilderment, and especially the sublime. How is it possible—I had to contemplate—that I had learned none of this history in my postcolonial biology classrooms growing up in India, nor in my settler colonial biological education in the United States? How can I at once know so much and yet so little about plants?

This book is my reckoning with the histories of science. In exploring the affective registers of my work, I focus on the idea of the sublime for two main reasons. First, I believe that the sublime and its associated values of transcendence, awe, beauty, glory, and grandeur best capture academic affective languages because they reveal the peculiar myths of the natural sciences. We *do* science not for the money or as a job but as something loftier: the dogged, relentless, and supreme pursuit of wonder, passion, curiosity, and "truth." In my interdisciplinary work, the vocabulary of the sublime has emerged repeatedly across various sites. Second, the sublime has an embrangled history with how "nature" and the role of the "human" in nature are theorized. The sublime evokes transcendence from material life, and immanence—something beyond the human, almost otherworldly. This idea of the sublime in the history of the natural sciences has shaped not only the gendered and raced prototype of the scientist but also theories in the natural sciences. The sublime, I argue, is a particular emotion that emerges through western thought and history and necessitates a split between nature and culture and between human and nonhuman. It permeates the epistemological foundations of research and knowledge-making practices.

A Botanical Journey: Four Tales

In my biology classes in both India and the United States, the history of botany was always presented as an established and incontrovertible account. The history was also always of a "modern" botany and celebrated through the works of a long list of great white men—Carl Linnaeus, Charles Darwin, John Ray, Jean-Baptiste Lamarck, Asa Gray, Charles Edwin Bessey, George Bentham, Joseph Hooker, and Frederick Clements, to name a few. But what if we put plants—not botany—at the center of analysis? That transforms our subject and allows us to understand how the "natural" world was theorized and systematized into what we call the modern science of botany. What was included and excluded in this science? Any systematic study of plants quickly leads us to the thoroughly embrangled colonial worlds of plants and humans and their coevolved histories, a topic absent from most botany classes but a central focus of this book. Historians of science have documented that where colonists went, they transformed landscapes through colonial forestry, plantation crops, and spice trades.[1] Colonial histories are fundamentally about the exploitation of colonized peoples' labor and of plants and other natural resources. For example, the histories of cotton and sugarcane cannot be told without a history of colonialism and slavery, and indeed vice versa.[2] The biologies of cotton and sugarcane shaped the labor policies and economic politics of plantations, knowledges that circulated and shaped colonial plantations across the globe. In such histories, it is impossible to extricate plants and plant agency as an abstract, universal, unique system, severed from the histories of other lives and the processes and practices of colonialism.[3]

In short, new insights from postcolonial and indigenous scholars remind us that modern botany can be taught as an abstract esoteric science only because the histories of colonization and slavery are excluded, as so many groups—black, indigenous, and people of color, disabled peoples, queer and trans communities, colonized and third world peoples, and women (except white women in the early years of botany, who were later marginalized if not forgotten)[4]—are excluded from the hallways of modern science. In putting plants at the center of analysis, we get a glimpse of how and why decolonizing projects in science are so enormously difficult and complex. It is not a tweak here and there but an epistemological and methodological revolution. As I immersed myself in these untold histories that have remained in the margins until recent years, I have come to reflect on my own botanical education. In hindsight, so much of my life and education makes sense now. The profound disconnection and deep sense of alienation that befell me, and that must indeed befall any student of botany from a colonized or enslaved nation, were not about me at all. Understanding the profound consequences of this whitewashed history is critical if we are serious about questions of inclusion, equity, and diversity. Until we learn to question the narratives we have been told about the past—to ask *what* these stories are, *who* these stories include and exclude, and *why* these emerged as *the* stories of botany—we cannot reckon with the violence of colonialism and its afterlives. Here I reflect on my own life and education as a case in point, one such journey narrated through four tales.

"OTHER" BOTANICAL IMAGINATIONS

I was born and grew up in India and benefited from a postcolonial education from elementary school through to my undergraduate years. When I refer to a postcolonial education, I am primarily discussing the system in India, but one that is in fact prevalent across the former British colonies. Given the multiplicity of languages in India and the divisive politics of elevating any single Indian language as the national language, India has many official languages. English often emerges as a "neutral" choice. Most urban middle-class schools like the ones I attended had English as the primary language of instruction. Indian postcolonial education followed a British style education, which meant that we specialized early. The textbooks were also deeply Anglocentric. My textbooks in history, literature, poetry, geography, mathematics, and the natural and physical sciences mostly covered well-accepted western theorists and disciplinary frameworks. While there was an occasional Indian scientist, like the Nobel laureate physicist C. V. Raman or the famed mathematician Ramanujan, the intellectuals featured in the textbooks were all from the west.

Since childhood, I had been fascinated with the natural world around me. So in high school, when I had to choose between the sciences, commerce, or the arts, I chose the sciences as my focus. After high school, when I had to pick a specialization in the sciences for my undergraduate years, I chose biology. I want to stress again how specialized this education was: during my undergraduate years, other than a study of language, I had no courses in the social sciences or humanities in my curriculum. An undergraduate degree in biology also meant no mathematics, a profound gap that I would need to repair in graduate school.

In my travels to the west, and through feminist and postcolonial studies, I have come to rethink and reexamine my early life. In India, I was immersed in a world of stories. Parents, grandparents, other relatives, and friends would tell and retell stories, especially from the epics *Ramayana* and *Mahābhārata*. In both epics there are innumerable characters, and over the centuries, stories have multiplied as each character becomes the center of a new story, completely transforming the unfolding of the epic. As a result, storytelling possibilities were endless. This was a magical, vibrant, colorful, and enchanted world. Western boundaries of life/nonlife, plant/animal, human/nonhuman, nature/culture, good/evil, and sacred/profane were largely absent. Gods would descend onto earth in numerous avatars, taking the shape of animals, humans, and animal-human hybrids. Humans could transform into animals, plants, and even rocks and then be transformed back again. Plants would whisper to each other and to other creatures around them. These imaginative possibilities played an active part in the plots of my childhood imagination.

Plants—their roots, shoots, leaves, and flowers—were woven into wonderful stories. In these fantastical stories, plants were never immobile, unintelligent beings but rather were active agents that built community and enabled narrative plots with the surrounding world. Plants housed transmigrated souls of the divine, human, and other sentient beings. For example, in these tales, the sweet fragrant jasmine emerged with the churning of the Ocean of Milk.[5] The tamarind tree, ever delicious for its sour fruit, was always home for myriad spirits. Despite its considerable shade, one was advised not to rest under it. The banana tree's short life is entangled in tales of matrimony and children. The banyan tree, with its expansive aerial roots, symbolizes all three gods of the Hindu trinity—Vishnu, the bark; Brahma, the roots; and Shiva, the branches. Our childhood was enraptured with the stories of Vikram and Vetaal: King Vikramaditya and the ghost spirit Vetaal, who lived in a peepal tree at the edge of a crematorium. In these stories, the ghostly spirit would continually outwit the king in clever and conniving plots. The sesame tree, we learned, emerged when drops of Lord Vishnu's sweat fell on Earth. The kalpa vriksha, or tree of life, was believed to grant your wishes. The peepal tree (sacred fig) and the banana tree are considered husband and wife if they grow together. The peepal leaf, with its symmetrical heart shape, made a delicate yet hardy surface for painting. I have spent many a summer drying the leaf, waiting for its skeletal structure to emerge, and then painting on it. Most minor medical symptoms of discomfort were promptly attended to by home remedies concocted of herbs and spices. Plants were integrated into life, essential and ubiquitous: as digestives, as beauty remedies, as games, and as musical instruments. I acquired this enchanted vision from these stories despite being raised as an urban child.

What is striking to me as I look back is that while I would call my childhood imagination magical and enchanted, it was never sublime, a concept I elaborate later. There was nothing otherworldly in these stories; all characters inhabited the same worlds. In my enchanted Indian imagination, gods, demons, plants, animals, humans, and even inanimate objects like stones were all interconnected, with varying but shared agencies. They could communicate with each other, and indeed in some stories each could turn into another (such as in the *Ramayana*, where the human Ahalya is turned into stone and then back into a human). Gods and demons are also not the binary good/evil that populate the western imagination. Gods can be bad and can do wrong, and demons can be pious and benevolent.[6] The notion of the sublime conjures a pure Eden that cleaves a false binary of separation between nature and culture. Thus, while Adam and Eve were banished into a harsh wilderness, Rama and Sita in the *Ramayana* were banished into an enchanted forest filled with talking animals and plants. They were at home and in harmony with the world around them. There is good and evil, but benevolence is not limited to the gods, nor sentience to the human. As I argue later, the sublime fundamentally implies the ability to transcend the human into an otherworldliness, but this otherworldliness was never in evidence in these mythological stories, and indeed seems incommensurable.

Once I entered school, we were inducted into a postcolonial education system where the legacies of the British still endure. As I grew up, my Indian upbringing was cast in a new light. Most significantly, distinctions between the real, the fictional, and the fantastical clearly came to be demarcated in my classrooms. I came to understand that the plant worlds I grew up with are in reality only fictional stories, superstitions rather than "real" knowledge. With time, my botanical education disenchanted my enchanted childhood imagination of nature. The scientific Latin names I had to learn by rote erased the luxurious and magical contexts of my childhood. They severed plants from the world I lived and knew. Common names that described plants were transformed into alien creatures. The beautiful "flame of the forest," which put out brilliant orange flowers that made trees seem like they were on fire, were now rendered *Caesalpinia regia*; the touch-me-not plant, which shriveled up if you touched it, was now *Mimosa pudica*; the banyan tree, the national tree of India, was now *Ficus benghalenssi*; the peepal tree, the sacred fig, was now *Ficus religiosa*; and the tall, lanky Asoka tree was now *Saraca indica*; and so it went. Suddenly, the plant world turned into Latin! While the Indian names always described the plant—as aflame or not to be touched—the scientific names were completely alien. This added another layer of rote learning in an education that already stressed it. I am sad to say that I was a rather compliant student and took to learning this new terminology as another chore that needed to be endured. Because my family moved around a lot during my childhood, the transformation of language was never complete, and I continued to nimbly code-switch between the vocabularies of my classes, my multilingual friends, and the local languages.

Thus, like other students of botany, I had to juggle multiple vocabularies and worlds. Most of us could name plants in multiple languages, and we had to add Latin to the many local forms. But learning the scientific term in Latin was enforced, and we were assured that because it was scientific, it was universal. There was a cosmopolitanism in our education in general, and an enforced cosmopolitanism in the convention of Latin botanical terms. One brought us into a scientific knowledge of difference across place, and the other flattened those very differences.[7] Scientists across the world used the same terminology. This was poor comfort for a child who didn't imagine traveling across the globe. Universality is most important to those how want to impose it, those who travel and have access to other lands and knowledge systems.

In my Indian botanical education, did we learn about botany and the spice trade, the political economy of the East India Company, and centuries of British colonialism and how they devasted and (under)developed India? Of course not. Did we learn about colonial forestry and the ways in which it transformed Indian landscapes? Again, no. Did we learn of plantation crops and how they transformed Indian agriculture? No! However, in our English classes, we were forced to recite poems like Wordsworth's "The Daffodils," Coleridge's "The Rime of the Ancient Mariner," and Shelley's "To a Skylark." For a population that had never seen (and were unlikely to ever see) a daffodil, albatross, or skylark, a postcolonial education was an education in alienation, where life on the ground diverged from life in the textbooks. In our spare time, childhood fiction was dominated by the world of a particular children's book author, Enid Blyton. We read of snow, strawberries, scones, crumpets, freckles, and any number of objects alien to our lives in India, and they consumed the poetry and prose of our literary worlds. No doubt, any good education should introduce new worlds and objects to students, but significantly, little from my lived experiences entered my textbooks. As Robin Kimmerer poignantly notes of her experience as an indigenous botanist in the United States:

> In moving from a childhood in the woods to the university I had unknowingly shifted between worldviews, from a natural history of experience, in which I knew plants as teachers and companions, to whom I was linked with mutual responsibility, into the realm of science. The questions scientists raised were not "Who are you? But "What is it?" No one asked plants, "What can you tell us? The primary question was, "How does it work?" The botany I was taught was reductionist, mechanistic, and

> strictly objective. Plants were reduced to objects; they were not subjects. The way botany was conceived and taught didn't seem to leave much room for a person who thought the way I did. The only way I could make sense of it was to conclude that the things I had always believed about plants must not be true after all.[8]

With time, life on the ground was a "third world" life filled with what seemed like the inconsequential, unimportant, and marginal, while the textbooks in our classes were the path to enlightenment, modernity, and science. Jamaica Kincaid channels this alienation beautifully in her novel *Lucy*.[9] Like me, Lucy was forced to learn and recite the poem "The Daffodils" and to learn to admire their beauty. After having now lived in a world with winters, I can appreciate the arrival of spring in ways I never could have in India. Daffodils herald the arrival of spring and the eruption of glorious landscapes of flowers and color. The poem is a joyous ode to the exhilaration that fields of daffodils fluttering and dancing in the breeze bring to the winter-weary soul. In India, we joked that there were three seasons—hot, hotter, hottest. The beauty of spring and daffodils was entirely lost on me. Lucy similarly grew up in the warm British West Indies, and like me she was forced to memorize the poem, but unlike me, she came to despise it. Forced to learn things about joys that were completely alien to one's own life was both alienating and annoying. In Kincaid's novel, Lucy hated the task immensely and resisted it, and in a hilarious moment she violently responds:

> I wanted to kill them. I wished that I had an enormous scythe; I would just walk down the path, dragging it alongside me, and I would cut these flowers down at the place where they emerged from the ground. (29)

Alas, I, as a dutiful postcolonial subject, took in the world as it was presented to me by my education. It is only through feminist and postcolonial studies that I have come to understand and appreciate the violence of colonialism's legacies, which remain enshrined in our knowledge systems. Decades later, my reeducation in the history of colonial science in India has evoked the outrage that Lucy feels and that Kincaid's powerful work evokes.

School in India was mostly enjoyable for me. In middle school in particular, I had two exceptional biology teachers, Mrs. Gowri Prasad and Mrs. Satyabhama, who inspired a fascination for the subject. If I asked them a question, they responded with several more. I read copiously, fueling my imagination. I feel I owe these two wonderful teachers my love for the biological sciences. Our school was in a narrow urban space, a tall white building with multiple floors next to a strip of sand where we could play games. There was little greenery that anyone could call nature. There was so little ground that for any and all sports, the school had to rent a nearby playground—not atypical of urban schools. So biological knowledge remained very much in the life of the mind.

Television came late to India. I was well into high school before we could afford a television set. In the early years, broadcasts were restricted to a few hours in the evenings, with more on the weekends. While there was some local content, many shows from the UK and the US were also popular. Three shows in particular captured my young mind—*Life on Earth*, *Cosmos*, and *Star Trek*. Of the three, David Attenborough's *Life on Earth* made the most indelible impression on me and is the one most relevant here. Attenborough, brother of the movie director Sir Richard Attenborough, is an English broadcaster who, in conjunction with the BBC Natural History Unit, was one of the early figures to present natural history to a television audience.[10] *Life on Earth* was just that—he traveled to the far corners of the planet and showed us how life flourished in the most unlikely and uninhabitable places—bottom of oceans, tops of mountains, dry deserts, salty marshes. The cinematography was stunning, and his narration, which he wrote himself, followed an easy and engaging storytelling style that coupled information with a friendly intimacy. His tone was passionate, infused with incredulity and reverence about the vast, bountiful natural world we had inherited. It would not be an understatement to say that you felt you were with him, sharing in the awe and wonder of his discoveries.

Attenborough's stories captured the inventiveness and wondrousness of the natural world. Highly popular, he followed *Life on Earth* with eight other natural history documentary series. Having recently read *On the Origin of Species*, I was besotted with Charles Darwin, and Attenborough appeared in my life at just the right time. Each week, he transported me to lands and spaces I had never seen or even imagined. Reading about plant adaptation is one thing, but watching the bizarre and beautiful innovations of plants and animals in vibrant color each week was incredible. In hindsight, this might have been my first experience with the sublime. Attenborough, a naturalist who had traveled the globe and studied the complexities of biology, modeled the sublime to perfection. His voice mixed wonder and authority and compelled you to marvel along with him at the extraordinary adaptations of life on earth. While my everyday urban Indian life involved a very particular and mundane engagement with nature, particularly the house lizard, the flying cockroach, and the house fly, he offered a profound sublimation with nature. Of course, I now realize that I was not alone. Attenborough spoke to a global audience of about five hundred million, reengaging urban populations across worlds with tales of natural environments that house most of the world's nonhuman inhabitants, lives that he brought from the corners of the globes into our living rooms.[11] I was enthralled, and I was hooked. I have vivid memories of watching the show with my father and having arguments about whether variation by natural selection could produce the beautiful, intricate, bizarre, and monstrous diversity on earth that graced our screens each week. As a staunch Darwin lover, I feel I defended him ably.

Attenborough also channeled the spirit of wilderness, and the importance of preserving and conserving life on earth was central to the spirit of his shows.[12] If he could communicate the perilousness of so many habitats, perhaps he could move us all to follow a path of preserving our natural resources. His most recent series presents the urgency of climate change and the need for action.

Attenborough was prone to tell "just-so stories," a term in evolutionary biology that refers to untested teleological explanations for biological traits and behaviors. But Attenborough revealed a world of evolution in action—there was a reason for everything in the world. The bat echolocates because . . . , the orchid flower looks like a moth because . . . To my young mind these stories were powerful, cementing my conviction in the power of evolution by natural selection. Every odd or wondrous part of a plant was an adaptation. He often presented these in a tone of tremulous wonder. There was always order, wisdom, and efficiency in the universe, and it was the job of scientists and naturalists to discover it. I was sold on the job!

Carl Sagan did for the universe what Attenborough did for the planet. Each week he presented the complexities of the universe, and with his infamous phrases such as "millions and millions" and "billions and billions," he reinforced the utter insignificance of humans. Transported to the outer reaches of the planet with Attenborough and the galaxy with Sagan, I was filled with wonder and awe. Both shows powerfully evoked the sublime and reinforced the insignificance of humanity in the face of a resplendent and innovative planet and universe. When I think back, I can still feel the youthful wonder and awe, and the emotions of those days are still palpable, shaping my desire to pursue a life in evolutionary biology. Of course, a life in science would change all that.

DEEP IN THE "CULTURE OF NO CULTURE"

I arrived from India to the United States for a PhD program in evolutionary biology/genetics.[13] The college in India where I completed my undergraduate degree had no experimental or research tradition, and biology lay largely in the realm of textbooks and practica where we had to perform prescribed activities like dissections, cataloging organisms, identifying parts under a microscope, creating herbarium sheets, and so on. None of this prepared me for graduate school. Even for a student with a research background, graduate school is a transformative experience—one enters as a student and leaves as a colleague to the faculty in the department. It is a profound transition to move from mentee to colleague. In a science program, one does that by learning to move seamlessly within the culture of science. Indeed, graduate school is a critical site where students are "enculturated" into a life in science.[14] For some it comes more easily than others. For those in the margins of science, this transition is difficult and alienating because of assumptions about which people make "ideal" scientists. Emerging from a Christian clerical tradition, the prototype of the scientist has remained remarkably enduring.[15] The famous "draw a scientist" exercise, where children are asked to draw a scientist, finds that children continue to represent scientists in very familiar tropes.

> The scientist is a man who wears a white coat and works in a laboratory. He is elderly or middle aged and wears glasses . . . he may wear a beard . . . he is surrounded by equipment: test tubes, bunsen burners, flasks and bot-

> tles, a jungle gym of blown glass tubes and weird machines with dials. . . , he writes neatly in black notebooks. . . . One day he may straighten up and shout: "I've found it! I've found it!" . . . Through his work people will have new and better products . . . he has to keep dangerous secrets . . . his work may be dangerous . . . he is always reading a book.[16]

This archetype of the scientist resonates with my experience. While the archetype is modulated in individual disciplines, some aspects remain remarkably stable—the assumption of scientist as male, white, older, unkempt, singularly devoted, passionate about his work, studious, arrogant, and bordering on the asocial.[17] A critical mode to prove one's mettle as a biologist is to enjoy endless hours of roaming and acutely observing the natural world. For a field biologist and ecologist, a love of nature and the natural world was paradigmatic. My postcolonial education and exclusively urban upbringing seemed to have ill prepared me for my graduate education. Nearly all my peers grew up deeply connected to nature; as children they spent hours in the woods behind their house, their families went camping in national parks, they hiked regularly, they traveled cross-country and sometimes went backpacking across the world during their high school or during a year off.[18] It was safe to say that I had never roamed a forest in my life. There were no woods near my urban housing, and the schools and colleges I attended had few resources for lengthy field trips. My love of nature came from books, television, and movies, not through personal and sensory encounters exploring the natural world. Ironically, it is my privilege as an urban middle-class and upper-caste Indian that engendered this alienation.[19] But in graduate school, and especially as an aspiring experimental field biologist, firsthand knowledge of the natural world was a critical asset. Field trips were ubiquitous, and not knowing the plants around me and not understanding local ecological communities or US forest succession redoubled a growing insecurity. Deeply ensconced within a graduate program, I came to be convinced that such a love of nature must indeed be a fundamental criterion for being a good biologist. My insecurity grew as my mind translated this into inadequacy. Only years after graduate school, as I have begun to read about the history of biology, have I come to recognize how central assumptions of gender, class, race, caste, nation, and urban/rural divides are in graduate school, and how such assumptions are intimately knit into professional expectations and identities. Working on this book, I have been moved by narratives of indigenous scientists and their love of, kinship to, and spiritual connections with the land and nature. In contrast, the spiritual worlds in postcolonial India have been thoroughly commodified, dismaying and distancing many like me.[20] Histories of slavery have also profoundly shaped the experiences of African Americans with respect to the land and nature.[21] Confronting histories of empire necessitates acknowledging and sustaining multiple relationships with nature.

As I grew alienated and felt increasingly inadequate, I considered leaving the sciences. I approached the women's studies program and was introduced to feminist STS. It saved my life! Feminist scholarship gave me the tools to understand and interpret scientific culture—the "culture of no culture"[22]—and through feminist STS, I began to understand what had happened, and this understanding profoundly shaped my academic trajectory. My entry into the world of feminist STS was important, and I began to deploy my newfound knowledge along with some friends. First, some of us in the department began a discussion group on life in graduate school. Subsequently, Rebecca Dunn, Lynn Broaddus, and I surveyed graduate students in the botany and zoology departments.[23] What we found was that while we had all learned to feign confidence, deep within, everyone felt inadequate and felt like "imposters" who were going to be found out any minute. Peers who we thought were supremely confident turned out to be profoundly insecure. These sessions were mind-blowing for many of us. Emotions came pouring out. People talked about how alienating graduate school culture was. What was especially salient was how much time everyone spent dealing with their insecurities, semiparalysis, difficulty with completing work, and writer's block. The survey confirmed much of this. While the degree of insecurity was gendered (women reported a much more precipitous drop in self-confidence), nearly all graduate students felt an erosion of self-confidence in graduate school. For me this was immensely consoling. The problem was not me but the scientific culture I lived in. These insights allowed me to stay in science and finish my degree.[24] I realized that I loved science and could be good at it; it was the culture that felt so hostile. Now feminist STS had given me the tools to recognize it for what it was.

INTERDISCIPLINARY HAUNTINGS: BOTANIZING IN NATURECULTURES

The insights of feminist STS also helped me understand the teeming emotional undercurrents of scientific culture. One of the most trenchant analyses of scientific culture comes from the work of Sharon Traweek, especially her influential ethnography of high-energy physicists, *Beamtimes and Lifetimes*.[25] In this book, Traweek memorably describes the culture of science as a "culture of no culture." It might also be described as a culture bereft of a lexicon of affect.[26] What would a culture that eschewed emotions and embraced the idea of objective, value-free knowledge look like? What would a culture that denied the existence of culture look like? Traweek points to scientific culture as a singular example. Her work revealed the importance of recognizing that science did indeed have a culture and gave me the tools to begin exploring it. It allowed me to look beyond the bromides of meritocracy and objectivity that usually stifled discussions.

This insight leads to the second example that emerged in a project I worked on with my mentor and colleague Mary Wyer.[27] During this National Science Foundation–funded project, we involved faculty and students in an exercise that asked each to list the "unwritten rules" of graduate school. When students enter graduate school, they are given a handbook and a list of expectations. Were there any rules not written down but still enforced? While faculty had a great deal of trouble naming such rules, students came up with a copious list. They argued that a primary rule was "be just like your advisor!" The pressures of graduate school were singularly about conformity—a project of intellectual reproduction, indeed of replication. Students described great pressure to conform, and to fake it if they couldn't. The culture was hierarchical, and the system expected the hierarchy to be maintained. The primary mode of interaction, students argued, were modes of competition and confrontation; confrontation, they argued, implied competence. Indeed, our department was a singularly competitive and confrontational culture. The affective dimensions of scientific culture were particularly striking. For example, one of the rules said, "The following emotions, topics, and behaviors are not allowed: crying, insecurity, laughter, personal problems, complimenting others." Another: "Helping others detracts from your personal achievements." Students talked repeatedly about how "feminine" emotions and any evidence of a "home" life were frowned upon. Of course, students argued that they did not or could not always follow the rules, and as a result they had developed various dissembling behaviors. They were not in the lab at all odd hours, but they would make it appear like they were. It was not that they did not cry, just that they learned to do it elsewhere. What emerged in this project reinforced the overflowing emotional undercurrents in the culture of science—undercurrents that could never be discussed. As Evelyn Fox Keller argues, the prototype of the scientist "poses a critical problem of identity: any scientist who is not a man walks a path bounded on one side by inauthenticity and on the other by subversion."[28] If you were "yourself" and did not subscribe to the stereotype, you were subversive, and if you faked it, you were inauthentic. There was no winning. Only a feminist project could reveal what lay below the surface.

But there was more. I realized that feminist scholarship helped me understand the culture not only of science but of knowledge-making itself. During graduate school, I worked with morning glories in order to understand the mechanisms that maintained flower color variation.[29] As with any experimental organism that one spends endless hours with, one cannot but develop an affective relationship with the organism. You must deal with the biology of the plant and the needs of your experiments. These come to shape the rhythms of your life and your everyday routines. As their name suggests, these plants are glorious in the mornings, and seeing them is always an incredible way to start the day. During my field season in the summer, I would wake up bleary-eyed and drive to my field site. I can still remember the spectacular sight of morning glories on a warm summer morning. Splashes of purple, white, blues, and pinks in intense, dark, and light hues all intertwined. The flowers were fresh, open, and shimmering in the morning dew. The bees and other insects buzzed around them. By noon, the flowers withered under the blistering sun. Perhaps the bees had done their magic or perhaps the plants had fertilized themselves. By afternoon there was virtually no color in the field; by evening it was entirely a field of green. The following morning, another glorious vision.

There were also annoying aspects to the plant. They were climbers, and so their tendrils were apt to seek out and climb over neighboring structures and plant. Go away for a few days and you had an entangled morning glory mess! It would take hours to untangle the plants from each other and retrain them onto their individual stakes. As an evolutionary biologist, I had to keep the plants separate since I needed to measure each plant and keep track of its flowers and seeds. The activity of gently disentangling the delicate tips to train them around their stakes taught me immense patience. The last thing I wanted was to damage the plant and lose a precious data point in my experiments.

Fieldwork is incredibly solitary. Most of the time, you are alone in your field site and cannot but develop complex affective relationships with the site and model organisms. Despite the intense heat of southern summers, and even after years of backbreaking and sometimes futile fieldwork, the morning vision of morning glories could always stop my heart. They made the mornings quite glorious, indeed endlessly sublime. Such descriptions are ubiquitous in experimental biology. Spending hours and days with experimental organisms makes one develop, in Evelyn Fox Keller's words in her description of Barbara McClintock, "a feeling for the organism."[30] In *Braiding Sweetgrass*, ecologist and indigenous scholar Robin Kimmerer captures this well.

> I've never met an ecologist who came to the field for the love of data or for the wonder of a p-value. . . . Science can be a way of forming intimacy and respect with other species that is rivaled only by the observations of traditional knowledge holders. It can be a path to kinship.
>
> These too are my people. Heart-driven scientists whose notebooks, smudged with salt marsh mud and filled with columns of numbers, are love letters to salmon. In their own way, they are lighting the beacon for salmon, to call them back home.[31]

The joys and passion for science that are ubiquitous in scientific memoirs, biographies, and hallway chatter are systematically erased from the scientific paper and record. This erasure helps sustain the patina of objectivity that shrouds scientific work. The scientific paper remains a strictly proscribed genre with clearly demarcated sections—introduction, materials and methods, results, discussion, conclusion—where the "doing" of science is limited to the materials and methods, with little by way of the unexpected occurrences that often are the "real" story. So much of the "enculturation" during graduate school was really an education in the social mores of scientific research.[32] In a "culture of no culture," where there is no vocabulary for the affective dimensions, it was difficult to process the joys and challenges of scientific research. The endless insecurity of whether you were good enough compounded the challenges of fieldwork, uncooperative research subjects, or unsuccessful experiments. Experiments can fail through no fault of your own. For example, once during a particularly hot summer, the vast majority of my genetic crosses in the greenhouse failed, so I did not have an adequate sample size for a robust experiment. The next year, the crosses worked beautifully. Thrilled, I planted hundreds of morning glory plants in the field. I returned the next day to find that over half of them had been eaten by cutworms. The cutworms did not even chew through the plants, merely snipped them an inch from the bottom, leaving stubs with the top fallen to the ground on the side. Processing the grief and fury of such frustrating moments is ubiquitous in experimental biology. I had spent a year in complex genetic crosses to produce the experimental plants, and then overnight I had lost the entire year's work. Nothing prepares you for that. These are lessons that are learned on the job. The *doing* of science in a "culture of no culture" means that each generation of scientists must learn the same lessons anew and must learn to thrive, cope, or leave the field's demanding silences.[33]

Wilderness: God, Nature, and Botany

I ground this essay around the sublime because I find it a particularly poignant node in my experiences with the natural and scientific. Understanding the sublime is a useful fulcrum in any project to decolonize botany. Even though scientific culture may claim to be a "culture of no culture," affective dimensions of research are real and everywhere—we just don't talk about it. As Robin Kimmerer's evocative quote above captures, the love of plants, the appreciation of beauty, and the exuberance in the face of the sublime are ubiquitous in science. I have also been struck by how my Indian childhood, while joyous and beauteous, was never sublime in the same sense of the world. It inhabited a different conceptual schematic of nature and the sacred. Here I explore the idea of the sublime and its genealogy within western science.

The sublime is derived from the Latin *sublimaire*, which links *sub* (up to) and *limen* (top of a door or threshold). The *Merriam-Webster Dictionary* defines the verb as to elevate or exalt, especially in dignity or honor; to render finer (as in purity or excellence); and to convert (something inferior) into something of higher worth. The sublime is more than mere emotion; it transcends the prosaic and mundane of everyday life. It evokes the profound, something just beyond human comprehension; it captures something sacred, otherworldly. As Philip Shaw explains, the sublime can be a whole range of items; a building or mountain may be sublime, as can thoughts, heroic deeds, or modes of expression. A grand revolution, noble ideals, and even the apocalypse may be associated with the sublime. Powerful words, music, aesthetics, or simply ideas like infinity can all seem beyond words and can fill us with a sense of the sublime.[34] The effect of the sublime may be "crushing or engulfing, so powerful that we cannot resist it."[35] It marks the limits of reason and expression, gesturing to what might lie beyond these very human limits. As Shaw astutely argues, *sublimity* refers to moments when our ability to understand or express a thought or sensation is defeated. Indeed, it is through defeat that the mind apprehends that which lies beyond thought and language. The sublime itself has a long history, and what is considered sublime has transformed with time as aesthetic, moral, cultural, and political mores have shifted.[36] That which is sublime can also been killed off and then resurrected.[37]

At its etymological heart, the term "carries the long history of the relationship between humans and those aspects of their world that excite in them particular emotions, powerful enough to evoke transcendence, shock, awe, and terror."[38] Nature is one prominent site in the history of the sublime.[39] Indeed, the debate on the relationship between humans and nature has been foundational in western philosophical thought.[40] The idea of a "pure" nature, untrammeled and unpolluted by the influence or presence of humans, is captured in the idea of "wilderness." In his iconic essay "The Trouble with Wilderness," environmental historian William Cronon traces the idea of wilderness in western thought. He argues that the feelings evoked by wilderness are entirely a cultural invention and contends that 250 years ago in America and Europe, it was rare for anyone to travel to remote areas of the planet to look for what we now understand as the "wilderness experience." Indeed, as late as the eighteenth century, the common use of the term *wilderness* was in reference to landscapes that were seen as deserted, savage, desolate, and barren—that is, wastelands. When Adam and Eve were driven from the garden, they entered a wilderness, a place that one came to only against one's will.

However, by the end of the nineteenth century, this perspective had changed. In 1862 American naturalist and philosopher Henry David Thoreau declared wildness to be the preservation of the world. In 1869 John Muir likened the Sierra Nevada to heaven itself. The American environmental imagination began demarcating areas of wilderness as preservation lands: Niagara Falls, the Catskills, the Adirondacks, Yosemite, Yellowstone. These early conservationists ushered in an era where "human development" was to be fought tooth and nail, while a supposedly untrammeled nature or wilderness was to be preserved.

Settler colonialism and westward imperial expansion are central to the US imagination of nature. Cronon ascribes the sources of this transformation to two foundational ideas: the sublime and the frontier. He argues that "of the two, the sublime is the older and more pervasive cultural construct, being one of the most important expressions of that broad transatlantic movement we today label as romanticism; the frontier is more peculiarly American, though it too had its European antecedents and parallels."[41] He also states that "By the 18th century this sense of the wilderness as a landscape where the supernatural lay just beneath the surface was expressed in the doctrine of the *sublime*."[42] Sublime landscapes were those "rare places" where we might "glimpse the face of God." "God was," he wrote, "on the mountaintop, in the chasm, in the waterfall, in the thundercloud, in the rainbow, in the sunset." In the work of romantics, the best proof that one had entered sublime landscapes was the unequivocal emotions they evoked, "nothing less than a religious experience."[43] The legacy of famed naturalist John Muir and other conservationists transformed the once frightening wilderness, the place of Satanic temptation, into a sacred temple, and as Cronon argues, this "continues to be for those who love it today."[44]

In the second half of the nineteenth century, the sublime awe of Henry Thoreau and William Wordsworth was domesticated through tourism, and the sublime emerged as central to the American imaginary—a centrality that has persisted into the present. For example, Leo Marx in *Machine in the Garden* likewise evocatively describes how the technological sublime "arises from an intoxicated feeling of unlimited possibility."[45] Modern biotechnology continues to channel this euphoria, and the sublime continues to be a central part of the scientific imagination.

This is a very abbreviated summary of a vast literature on the history of the environmental and natural sciences. As philosopher Michael Marder argues, the development of western philosophy is transformed by the growth of a sublime botany, where "philosophy itself changes beyond recognition: *philo-sophia*, the love of wisdom, is brought in life with the help of *phyto-philia*, the love of plants."[46] Through these philosophies, we can trace the deep entanglements of our conceptions of nature, human, humanity, and the religious and sacred. The idea of the sublime is key to capturing a long-enduring and critical binary of nature and culture, human and nonhuman, where humans and culture are seen as outside of nature. Conservation biology and its pursuit of a "pure nature" through preservation and antidevelopment has cemented this binary as a foundational idea in the natural sciences. Indeed, it grounds academic distinctions between biological sciences (basic sciences) and environmental sciences (applied sciences). Thoreau encapsulates this transformation in a simple sentence: "Once I was part and parcel of Nature; now I am observant of her."[47]

Beyond the binary, Thoreau captures another important thread prevalent in feminist thought: that like nation, *nature* is always rendered feminine and racialized and thus always rendered in need of mastery and control. The histories of colonialism teach us that botany is simultaneously a gendered and racialized project. Starting with Carolyn Merchant's *Death of Nature*, feminist scholars have chronicled how much western colonial expansion, industrialization, and development relied on unrestrained commercial exploitation of natural landscapes.[48] Environmental historians have argued that we should understand imperialism as fundamentally an ecological project in which humans, plants, and other species were shuffled around the earth in schemes for colonization and conservation.[49] The science behind the imperial ecological project is what emerges as fields, such as botany and conservation biology. The gendered logic of feminine nature is always racialized and also imbued with colonial logics of dominance over nature. In the name of civilizing the native, not only can native lands be colonized and exploited, but so must native peoples. An important part of colonization projects was to frame native lands as wild and unsafe and peoples as nonexistent or savage, thus opening up the need for military incursion.[50] Native knowledge and knowledge-making practices were delegitimized, erased, expunged. In India, for example, the British Raj helped create a new class of subaltern who were unculturated and unable to interface with the precolonial past, yet unable to be fully sublimated into the metropolis.[51]

Given this history, it should come as no surprise that the science of botany emerged as big science and big business, critical to Europe's ambition as a colonial trader in tea, spices, and other plant-based goods.[52] Colonialism ushered in a massive reshuffling of global biota. One can and should understand the botanical sciences as a significant legacy in the afterlives of empire. Colonialism fueled an extractive economy through the objectification and commodification of the colonized world and the destruction of local infrastructures and knowledges.[53] The term *extraction* captures capitalist modes of economic exploitation.[54] In its place, the colonial order installed the biological sciences as the universal, abstract, and expert knowledge.

Local ecologies were transformed by colonial logics.[55] Colonial routes of plants included the famous spice routes and transportation of lumber resources, plantation crops, and horticultural specialties but also a trove of other agricultural plants and animals. Horticultural societies and gardens across the west cultivated the exotic and curious. Kew Gardens, the hub of the natural sciences, and other such sites became repositories of the world's biota.[56] I take you through this brief excursion into the history and philosophy of the conceptual terrain of feminist STS because it is critical to recognize why botanical theories and frameworks are not just a mirror to the plant world but rather a colonial conceptual apparatus for the justification of imperial conquest and expansion. The sublime is central to this story. A focus on botany helps us understand how colonialism encompassed worlds beyond the human into plant worlds. To decolonize botany is to dismantle this immense apparatus of western epistemology, methodology, and methods. The sublime in particular reveals a crucial distinction between the modern and nonmodern at the heart of conceptual differences between the western and nonwestern. For western science, plants were a resource to be named, systematized, studied, and then efficiently extracted for colonial wealth. In contrast, as I recounted my early life in postcolonial India, the binaries that frame western science were never prominent. In Indian mythologies and many nonwestern cosmologies, objects are relational, not hierarchical. Plants are not lesser than animals, humans, or the nonliving but infinitely entangled and interchangeable. It is the binary formations of nature/culture, human/nonhuman, and living/nonliving that ground the "sublime" of western ideas of nature that allowed humans to feel the sacred and the divine.

Conclusion: The Embrangled, Enchanted Worlds of Plants

The best moments of interdisciplinary revelations for me have been entirely fortuitous. It began in graduate school in evolutionary biology when Mary Wyer, a mentor and colleague in women's studies, asked me about my work on morning glories. She listened patiently as I described my work in understanding the evolutionary maintenance of flower color. When I finished, she looked at me with what seemed like perfect comprehension and remarked that yes, she understood that I worked on diversity! Until that moment, my work on "diversity" lay in gender studies, not in the fields of evolutionary biology. Likewise, *variation* was a term in my world in evolutionary biology but not in gender studies. It was exhilarating to make the connection between biology and feminist studies, and my research has never been the same. The ensuing book *Ghost Stories for Darwin* traces my growing recognition of how cultural ideas of diversity are intimately connected to the related theories of "variation" in evolutionary biology. Understanding the foundational role of gender and race in the making of science was revelatory. Work on my next book, *Holy Science,* challenged my understanding of science as purely a western enterprise and forced me to contend with the powerful impact of colonialism on our landscapes and knowledge systems. Science did not just emerge alongside colonialism in a separate sphere. Rather, science as we know developed for and in aid of empire; there is a coloniality to science. The particular version of gender and race as we discuss them in science are themselves colonial constructs. Colonialism is thus at the heart of the enterprise of science. Together, my work on both books has helped shape this one. I now recognize that science is always conducted within cultural and political contexts and that one can trace the circulations of theories across the worlds of nature and culture. Ever since, I have refused the false choice between botany and feminism. I should not have to choose since they are not separate but always co-constituted. For me, working with one always means working with the other.

Interdisciplinary work has been filled with such moments. Disciplinary silos teach us not only *to see* but, more significantly, *not* to see. Interdisciplinary work involves learning *and* unlearning. In feminist STS, nature and culture are not ontologically or epistemologically different or distinct projects but best understood in their imbricated histories as *naturecultures*.[57] In my recent work on invasive species, for example, when I look at fields of plants, I do not see just plants. I see creatures whose social, historical, and biological lives have shaped their evolutionary pasts and futures. Their embrangled histories with humans have rendered them native and foreign, desirable and undesirable, threats and saviors. I understand my interdisciplinary location in a women, gender, sexuality studies (WGSS) department rather than a biology department as a product of the colonial histories of the biological sciences, legacies we have scarcely come to reckon with. Projects that engage critically with race, gender, and colonialism still occupy the margins of disciplines, and WGSS, ethnic studies, and queer studies programs continue to remain in the margins of academe. My journeys through disciplinary and interdisciplinary work have convinced me that it is critical to liberate ourselves—intellectually and professionally—from the silos of disciplines. It is in the margins, in the interstices, that we find the tools and freedom to imagine knowledge anew. These margins free our senses and our minds to think, feel, and express the art and science of knowledge-making. In tracing and untangling the colonial, postcolonial, and neocolonial histories of naturecultures, we open up botanical histories, theories, and methodologies to glorious new interdisciplinary explorations. I hope that one day botanists and plant lovers can together embark on a project of a collective botanical reimagination.

CHAPTER TWO

The Coloniality of Botany

Reckonings with the History of Science

Colonialism is not content to impose its rule upon the colonized country's present and the future. Colonialism is not satisfied merely with snaring the people in its net or with draining the colonized brain of any form or substance. With a kind of perverted logic, it turns its attention to the past of the colonized people and distorts it, disfigures it and destroys it.

FRANTZ FANON | *The Wretched of the Earth*

Dreaming, even in inclusive and multicultural tones, of developing an ideal settler state implicitly supports the elimination of Indigenous peoples from this place. . . . And for those of us who remain, our intimate relations with these lands and waters continue to be undercut and our memories relentlessly erased when the extractive nation-state continues to be dreamed.

KIM TALLBEAR | "Caretaking Relations, Not American Dreaming"

The failure of academic feminists to recognize difference as a crucial strength is a failure to reach beyond the first patriarchal lesson. In our world, divide and conquer must become define and empower.

AUDRE LORDE | *Sister Outsider*

A DELICATE VIOLET LIES PRESSED IN A BOOK OF SONNETS. A translucent rose graces a love note. A sprig of lavender scents a fragrant candle. A dried daisy glows on an iridescent luminary. A wisteria branch lies pressed, displaying its cascading inflorescence on an herbarium sheet. Pressed plants have a long history; they span the artistic, popular, representational, sartorial, sensual, and scientific worlds. In the biological sciences, animals are pickled and plants are pressed. "Pressed" plant specimens emerged as a crucial method of scientific identification and classification. Herbaria, collections of plant specimens, were critical for developing the scientific, systematic, and universal codes deemed necessary for the easy identification of flora across the globe, crucial for the extractive logics of empire.

The site of the herbarium cataloging the richness of plant diversity vividly captures the colonial project—a global mortuary that revels in the violent necropolitics of botany. As the botanical sciences developed, herbaria and herbarium samples proliferated across the globe, becoming cemented as central repositories of a botanical imaginary of a global, universal science. Today they endure as colonial sites largely housed in the west, still controlling botanical norms. The oldest surviving herbarium is credited to Italian artist and herbalist Gherardo Cibo and his contemporaries in the mid-sixteenth century.[1] Dead remnants of plants on herbarium sheets are an apt metaphor for the epistemologies of empire, yet impoverished sites for telling the story of vegetal life on earth. Colonial architectures of knowledge claim universal and singular knowledge—a singular science, epistemology, and methodology—and botany as a study of plants without people.

The Stories We Tell about Botany

The "epistemic story of imperialism," Gayatri Spivak writes, is "the story of a series of interruptions, a repeated tearing of time that cannot be sutured."[2] And yet, as Priya Satia argues in *Time's Monster*, these torn fragments of time were sutured into a triumphal, linear, and progress-oriented vision of European colonialism *as* history. She compels us to ask what it would look like for the field of history to truly reckon with its past. How could we make history anew?[3] She argues that the reckoning is not about progress itself but rather a recovery from the "imaginary of progress." The problem is the discipline itself—how academic thought in nineteenth-century Europe helped make empire by making the project of colonialism ethically thinkable, and how, in turn, empire made and remade the historical discipline. The ethical tools that colonists developed bypassed their conscience even as they unleashed murderous violence on "backward" peoples across the planet. "The story of empire," she writes, "is one in which the villains of history, if there are any, are historians themselves."[4]

Satia's argument is a story of all disciplines, including the botanical or plant sciences. As she argues, while the field of history took shape among a tiny group of elite Europeans, their influence was considerable. Grounded in the conviction of their absolute power, indeed a divinely ordained destiny, they endowed themselves, and only themselves, with historical agency. As a result, history was theirs to tell. In making the field of history, they valorized not only the project of colonialism but also their role in it. "All other beings—animals, landscapes, plants, and of course, the vast majority of humans—were, in their view, brutishly wedded to 'Nature' and were therefore incapable of historical agency. In other words, built into the foundations of History, and indeed, many other disciplines in the Humanities, is the repression of some of the most important questions about human existence on this planet."[5] This deanimation of the natural world is key to how the biological sciences would come to exploit the natural resources of the colonies and remake knowledge through the eyes of the west. The dried and dead plants in herbarium sheets are a striking case in point. It is through histories of indigenous, postcolonial, and queer studies that we now see the depravity of the term *humanism* and come to recenter and appreciate the rich traditions of peoples excluded from this fabled "history"—indigenous, colonized, queer, trans, poor, and disabled peoples, and of course nonhuman inhabitants.

The Coloniality of Science

How to narrate the history of colonial botany, the ongoing colonial regimes of settler societies, and their continued presence in the many afterlives in the postcolonies? How to tell histories mobilized as a potent site for the management and recuperation of white, western patriarchy? Leanne Simpson answers succinctly,

> Really what the colonizers have always been trying to figure out is: "How do you extract natural resources from the land when the peoples whose territory you're on believe that those plant, animal and minerals have both spirit and therefore agency?" . . . You use gender violence to remove Indig-

enous peoples and their descendants from the land, you remove agency from the plant and animal worlds and you reposition *aki* (the land) as "natural resources" for the use and betterment of white people.[6]

Colonialism is fundamentally a story of violence. Violence at multiple levels: a genocide through a willful dehumanizing and killing of peoples, an ecocide through a deliberate destruction of the environment and life within, and an epistemicide, the destruction of existing knowledge as well as local and indigenous ways of knowing.[7] It is violence incarnate: a teeming, inventive life on earth decimated, cleaved into hierarchies of human and nonhuman, superior and inferior, master and slave, native and alien; a vital, vibrant planet rendered into desolate landscapes of cracked and crushed earth, an ecological apocalypse of pillaged lands and peoples through slavery and empire.

A feminist retelling of this story insists that there is no abstract Anthropocene (from Greek *anthropo*, human, and *cene*, new) but a sexually and racially stratified Anthropos—where some anthropos colonized, conquered, enslaved, and dehumanized other anthropos. Then they returned home with their rich spoils to build fortresses and enact borders and laws to keep and protect their stolen fortunes. No pristine Eden here, no purity, no corner untouched by colonialism and its afterlives and the rapacious tentacles of modernity.

Science is integral to this project. Through conceptual landscapes of gender, race, class, caste, indigeneity, sexuality, ability, and binaries of sex/gender, cis/trans, native/alien, black/white, heterosexual/homosexual, disabled/able-bodied, colonizer/colonized, and healthy/unhealthy, science has created biopolitical landscapes of inferior and superior bodies. These conceptual blueprints haunt biological landscapes, including plantscapes. For example, plant sex mirrors frameworks from human reproductive biology—so males sow their seeds widely, while females nurture their progeny. Contemporary sciences, in the name of rectifying the horrors of the past, have recolonized rather than decolonized. Through nativist politics of purity, fields such as conservation biology, invasion biology, restoration ecology, reproductive biology, plant systematics, and agricultural biotechnology, and sites such as museums, gardens, and seed and germplasm banks are all savior sciences and practices that preserve the legacies of empire.

It is best to not talk about colonialism in the singular. There are many colonialisms. Broadly, we can define *colonialism* as the conquest and control of other people's land and goods.[8] But viewed from the long arc of history, while colonialism precedes the expansion of European powers into Asia, Africa, and the Americas from the sixteenth century onward, it was European colonialism that ushered in a level of violence that was unprecedented.[9] These more recent colonial regimes profoundly shaped the globe, the environment, plant lives, and the histories of botany. For the purposes of this book, it is this recent period—the expansion of European powers starting in the sixteenth century—that is most relevant. European colonialism was not singular either. Multiple nations colonized the world—Belgium, Britain, Denmark, the Netherlands, France, Germany, Italy, Portugal, and Spain. They varied in their ambitions, scope, practices, longevity, and impact on countries and the globe.

For botany, the histories of British colonialism are particularly salient. Four core patterns emerge. First, among recently independent nations, postcolonial elites often embrace and breathe new life into colonial legacies. In India, for example, postcolonial elites have strengthened the organizing principles of colonial forest administration, thereby continuing to curtail the rights of forest communities.[10] Decolonization here is no longer about colonialism per se but about dismantling oppressive structures (sometimes related to colonialism) that disenfranchise many marginalized groups. In India, anticolonialism has been cynically appropriated and subverted by Hindu nationalists to promote a Hindu supremacist nation. More than colonialism, it is nationalism that looms large. For this reason, I use the term *countercolonial* rather than *anticolonial*. Second, settler colonialism endures, especially in the United States, Canada, Australia, New Zealand, and South Africa.[11] Decolonization in this context means the rematriation of Indigenous land and life. While ideas, epistemologies, and ecologies are a rich part of the writings of indigenous scholars, stolen land looms large. The relationship of land to postcolonial migrants and forced migrants through slavery and indentured servitude are vastly different. Third, legacies of slavery and conquest endure in living bodies. Here, we need to track and address their ongoing violence.[12] For example, histories of slavery shape contemporary health inequalities.[13] Plantation and colonial crops enduringly transformed the economies and labor in colonized nations. Finally, across the globe grassroots movements, especially in Latin America, have fought to reclaim land and science to challenge knowledge from Eurocentric epistemes, conceiving and offering rich theories and practices of decoloniality.[14]

Colonialism thus has many geographies and histories. Understanding its immense and heterogeneous legacies is a complex endeavor. The complexity gets lost in the monolithic language of colonialism. For this reason, I have turned to the language of *coloniality*. The term comes from Peruvian sociologist Aníbal Quijano in the "coloniality of power,"[15] a concept developed in the context of the Americas. Quijano argues that the European colonial project in the sixteenth century ushered in the genocide of indigenous peoples and the organization of the African slave trade, and with those practices established a new global order that endures on to this day. The European colonial project is significant because it brought the eastern and western hemispheres together in a "single field of imagination" and the makings of a global imperial power.[16] Also, because Columbus landed in the Americas mistakenly thinking he was in India, the term *Indians* takes on a more universal note. *Coloniality* thus brings together the impact of the various colonialisms I outlined above. We can talk about the *coloniality of power*, the *coloniality of gender*, and through them the *coloniality of science and botany*. We see how long-standing patterns of colonial power shape infrastructures of institutions, including science and particularly botany.

As I follow colonial history forged across the globe, I mean to write, and I hope I am read, as in solidarity with scholars and activists who are working to undo colonialism's legacies across the globe. As the project of decolonization and decoloniality sweeps disciplines in the academy, it brings urgent questions to the fore, compelling academics to examine the past, to listen to voices long silenced and marginalized. It has arrived at the doors of the plant sciences. The decolonial turn is upon us; what do we do?

Decolonization and Its (Dis)contents

Audra Simpson reminds us that the state is not only repressive but also educative.[17] The state shapes what we consider to be common sense through apparatuses and ideologies (academia being an important one) that help normalize colonialism and its logics. Colonialism developed vast apparatuses that shaped relations between peoples, lands, nature, and civilization. Ushering in specializations, it imposed sexuality, legality, raciality, language, and religion in specific and particular ways.[18] As Frantz Fanon reminds us, "Decolonization, as we know, is a historical process . . . it cannot become intelligible nor clear to itself except in the exact measure that we can discern the movements which give it historical form and content."[19] Only by understanding history and the formations of botany can we approach any project of countercolonialism.

COLONIZATION AND ITS LEGACIES

If colonization was not an event but a structure, how do we address colonial regimes deeply sedimented into the structures and everyday practices of modern science? Language is central. Rethinking and reframing what we mean by terms is a first step, but the process cannot end there.

At the heart of epistemicide are the representational practices of the west and a politics of purity—pure nature (or wilderness) versus impure nature (anything tainted by humans). Such a binary immediately puts humans outside of nature. Similarly, the "great chain of being," premised on enduring hierarchies of race, gender, and nation, organizes biological imagination. Throughout this book we see the representational politics of plants—how plants come to become native or alien, to acquire sex and gender, to be named and renamed and placed in the tree of life. Any project of decolonizing must also acknowledge that western science, including botany, is not purely western. The circulations of science during colonial rule and the spread of knowledge from metropole to colony, and between colonies, created knowledge transfers and transnational circulations of scientific knowledge.[20]

COLONIZATIONS AND DECOLONIZATIONS

Decolonization spans a wide spectrum of ideas that include the politics of post-independence, restorative justice for indigenous communities in settler nations, and a trans-contextual understanding of deep-rooted structural impacts of the colonial way of thinking.[21] Academic fields and legal, political, cultural, and academic institutions have all begun an introspective journey about decolonization; how much is lip service only time will tell. Rolando Vazquez argues that the struggle to decolonize and undo colonial difference involves a threefold path that includes understanding several ideas: modernity's way of "worlding the world as artifice, as earthlessness"; how coloniality, in destroying relational worlds, helped "un-world" the world; and how decolonial thought can bring "radical hope for an ethical life with earth."[22]

In thinking about decolonization, it helps to consider settler colonialism and postcolonialism separately. Grounded in two separate literatures, the two have different temporal registers and political urgencies. Eli Nelson defines "settler science" as a cycle of termination, appropriation, and forgetting. Indigenous and decolonial thought echo through the histories of botany—colonial botany emerges through a process of termination and appropriation, and then those roots are forgotten as the "science" becomes sanctified as "pure." Such science, rooted in the appropriation and exploitation of indigeneity, works to annihilate Indigenous peoples.[23] Max Liboiron, in their powerful *Pollution Is Colonialism*, homes in on the key colonial logics:

> Under what conditions does managing, rather than eliminating, environmental pollution make sense? The answer, colonialism: the assumed entitlement to Indigenous lands for settler and colonial goals; the genocidal and other practices through which these lands are cleared of Indigenous people (and myriad other regulatory, fiscal, and criminal practices through which access is maintained); the scientific theories that operationalize settler and colonial entitlements to lakes, rivers, environments, atmospheres, and bodies as sinks for pollution and that eliminate other types of relations that threaten access; the epistemic frameworks that flatten and transform Land into Nature, Resource, and Property.[24]

Conceptually, indigenization represents a move to expand academic knowledge to include indigenous perspectives in transformative ways.[25] Decolonial indigenization insists that we reorient the imbalance of power relations between Indigenous peoples and the nation, thus transforming the academy into something dynamic and new.[26]

Nonwestern knowledge systems (including those present but marginalized in the west) offer alternate modes of theorizing and understanding the world. For example, unlike Aristotle's "great chain of being" that places humans at the pinnacle, Anishinaabe creation teachings place humans at the bottom of the four interconnected orders of life (first earth forces, then plants, nonhuman animals, and finally humans). This worldview reminds us that we must never forget our dependence on the planet's other life forms.[27] Indigenous perspectives on plants across the world are firmly rooted in relation to land and soil.[28] In a similar vein, Kim TallBear offers the relational cosmologies and epistemologies of the Dakota people.[29] In recent years, we have seen an explosion of work on indigenous ecologies, sciences, and knowledges.

But decolonization must not become recolonization. Decolonization must not mean appropriating indigenous knowledge as science.[30] Colonization was built on practices of appropriation, stolen land, and broken promises. As many Native scholars ask, Do we want sacred knowledges to be appropriated by academia? If they stole it once, might they not do it again? After all, science has commodified, bought, and sold knowledge for centuries, and "cultural restitution is a much larger process than just handing back a stolen artefact. You have to really deal with the issue and educate communities about decolonisation, which is more difficult and takes more time."[31]

Finally, TallBear cautions against fetishizing the indigenous.[32] Too often, the term *indigenous* has become consumed uncritically as a commodity, with little engagement with the rich and critical traditions of indigenous studies. As TallBear insists, authentic indigenous cultures cannot become a site of "static, generic, and uncomplicated traditionalism." Cultures are dynamic, and once fossilized, she argues, much is lost; it does not guarantee survival and in fact may hasten the demise of Native American peoples and cultures. A celebration of the sublime in botany is not for a romance of a pure or native past but in awe of difference and ways to reckon with it.

The postcolonial brings a different set of issues.[33] Just as indigeneity is not singular, neither is the postcolonial. It matters where you are. If you start from India, or Kenya or South Africa, or Brazil, the answers are as varied as colonial lives and afterlives. Even within a country like India, understanding the world from the point of view of a Dalit woman, a member of the Santhal tribe, a billionaire, or a poor upper-caste Brahmin will yield different views. Indeed, in a postcolonial country like India, decolonization is perhaps not the urgent question, because with the rise of Hindu nationalism, the vocabulary of anticolonialism and decolonialism have been thoroughly appropriated and then subverted by a Hindu supremacist state. The terrains of settler colonialism, postcolonialism, and decolonialism are complex and fraught. There is no singular solution for "decolonization." As Tania Pérez-Bustos warns us, a decolonial paradigm should not become a singular explanatory tool for the commitments and practices of diverse social justice struggles and radical thinkers across the globe.[34]

Maybe we need to decolonize decolonization! A central challenge in many postcolonial societies is that its elites, groomed by the colonizers, emerged as its new leaders. They have often reproduced the inequality of empire and structures of coloniality. As Yarimar Bonilla argues, "that's not a bug—the modernist project of decolonization has not failed, it is working as intended."[35] Decolonization cannot be just an exit of the colonizers, or "flag independence," but must include a fundamental undoing of colonial logics, moving beyond settler (and masculinist) logics of conquest.[36]

Terraforming the Planet

"Only once we imagined the world as dead could we dedicate ourselves to making it so," Ben Ehrenreich writes.[37] Colonialism erased local cultures and natures as it terraformed indigenous territories to erase local ecologies and their histories with the land. Ecological intervention was not an incidental effect of European colonialism but an explicit aim. The goal was "terraforming": turning parts of earth that Europeans considered wastelands into their definitions of productive land.[38] Early accounts of the colonial project describe Indigenous perceptions of Europeans as follows:

> They say we have come to this earth to destroy the world. They say . . . that we devour everything, we consume the earth, we redirect the rivers, we are never quiet, never at rest, but always run here and there, seeking gold and silver, never satisfied, and then we gamble with it, make war, kill each other, rob, swear, never say the truth, and have deprived them of their means of livelihood.[39]

As historian William Cronon notes, European perceptions of what constituted a proper use of the environment became a European ideology of conquest.[40] It is this violent legacy that confronts us in the tools and methods of botany. Indigenous ecologist Robin Wall Kimmerer evocatively contrasts the two dramatically different world views—Skywoman and Eve:

On one side of the world were people whose relationship with the living world was shaped by Skywoman, who created a garden for all. On the other side was another woman with a garden and a tree. But for tasting its fruit, she was banished from the garden. . . . That mother of men was made to wander in the wilderness and earn her bread by the sweat of her brow, not by filling her mouth with the sweet juicy fruits that bend the branches low. In order to eat, she was instructed to subdue the wilderness into which she was cast.[41]

Allegories across the world offer rich visions of relations of peoples, land, and life. Once out of the Linnaean labyrinth, we have many choices.

Fables for the Mis-Anthropocene: Chirp, Play, Love

"Noor! Noor! Where are you?" a voice cried.

It had been a long day in the hospital and, as usual, Salma's daughter was missing. Salma made her way to the gardens nearby. Noor was nowhere. She looked up, and sure enough, she saw Noor sitting at the edge of a branch, magnifying glass in hand, inspecting a mulberry tree.

"Noor, it is time for dinner. Come along," Salma ordered.

Noor climbed down the branches quickly. She was fearless, Salma thought, making her a little nervous but always proud. By now she was used to seeing her daughter's curiosity take her to unexpected places.

"I'm so hungry," said Noor. "I was watching the worms on the mulberry tree. They are so loud! Chomp, chomp, chomp, chomp—you can hear them! Have you ever heard them?"

"No," admitted Salma.

"I recorded them. I'll play it for you," Noor said. "They may seem small, but they have strong jaws! That is why I was using my magnifying glass. I can't wait to show it to my friends tomorrow." Noor chattered away as they walked home.

Noor was ravenous indeed, with the most curious of tastes. She took a piece of round pita bread on her plate and spooned atop it a large amount of lemony hummus. "My favorite," she said. To this she added some grilled tofu, broccoli, radish kimchi, raspberry jam, and finally some jalapeños. Salma winced, but Noor was her eclectic self. She wrapped the pita around and chomped in. "Yummy," she said, closing her eyes to savor it. "Spicy, salty, sweet, sour, and umami all at once in every bite."

Her pita pocket eaten, she washed her plate and used her allotted time on the computer to enter the Chirp-Net. According to Wikipedia, the Chirp-Net connects "a community of children across the world to share stories about plant and animal life. With time it emerged as an unparalleled network for showcasing children's knowledge-making abilities, creating a vibrant model of distributed participatory investigation." The name Chirp-Net was inspired by the vulnerable but eager cries of newly born chicks—*chirp, chirp*—a hopeful metaphor for the wails of a dying planet. On the Chirp-Net, children pledged to be friends with plants and animals that had no representation in governments, law, or industry.

As the planet approached the mid-twenty-first century, environmental problems were rampant. Powerful people in governments and corporations did nothing. Plants and animals were dying in large numbers; the air and water were heavily polluted. Health problems, especially of the young, were ubiquitous. Led by youthful activists across the globe, the world saw the eruption of the "Raucous Spring," a global revolution of young people who proclaimed, "Enough is enough!" They mobilized their parents, neighborhoods, towns, cities, and nations, holding sit-ins in households, schools, town halls, downtown squares, front lawns, and religious and civic institutions. With the rallying cry "The planet's shattered, start the chatter," they launched an unprecedented and successful movement. After all, supreme court justices, politicians, bankers, corporate heads, and managers had children; the rich and poor had children. With lively scripts, every child engaged their parents, relatives, friends, teachers, and neighbors.

At first, adults thought they would easily fend off these challenges. But they were surprised; the scripts were powerful. Next, some adults turned authoritarian. Children recorded many spankings, groundings, and other punishments. They responded with silent retreats and hunger strikes. Friends smuggled food and drink to them if necessary. Many radical ideas circulated through the Chirp-Net. In rich neighborhoods, students pledged to mix bottled water with polluted tap water from poorer towns, throwing parents into a panic. Was the water in their house safe? The water in the supermarket? If they'd had the resources, they would have piped polluted air and water into the courts and government buildings.

With time, children drew support from teenagers and adults of all ages from across the globe. The world had been waiting for this moment, it seemed. The movement gained momentum. A critical group of adults the world over believed in the children's cause. They wanted a future for their children.

Of course, not everyone was a fan. Some adults, mostly officials and media sources, called the crusade terrorism or guerrilla warfare. The world over, kids challenged politicians, the wealthy, and those in the highest echelons of power. In turn, the people in power tried to topple, subvert, and challenge the children. Politicians invoked laws and attempted arrests. But children made poor figures as antinational terrorists, especially when they represented such a large cross-section of the population. The solidarity of young people fighting for a livable future was unprecedented. History reminds us that revolutions are rarely predicted, but when they happen, we learn from then. Such was the case with the Raucous Spring.

The success of the Raucous Spring emboldened young people. Fighting a movement that nurtured children across the world and across differences proved impossible. Funny and ironic memes of phrases used by adults attempting to thwart the movement circulated, such as "in the real world," "it isn't as bad as you say," "if you persist in this insanity," "if those people became our neighbors," "if we became poor," and often, "you cannot." Social media was abuzz. The Raucous Spring mobilized a diverse, global alliance. Children, everywhere, wanted a future.

This gave birth to the Chirp-Net. Children concluded that the fundamental problem was that their "education" in the classroom was utterly removed from their lives. *We want to understand the world around us*, they said. *We want to learn how to solve real-world problems, to make the world a better place. We want an education that serves the people, not the powerful. Stop duping us*, they said. The world was nothing like what they were learning in their classrooms. And so emerged young investigative scientists—those who studied plants, animals, corporations, sports, schools, institutions, politicians, pollution, and so on. Chirp-Net emerged from the recognition that humans in positions of power were alienated from the world, from other humans, from plants and animals.

With the motto "Chirp, Play, Love," children met, and together they played with plants and animals. Friends and friendships were a key goal of the Chirp-Net. They met online, in the woods, gardens, and front yards, and sometimes at home, but the key point was to carefully observe the surrounding world. With microscopes, magnifying glasses, binoculars, or any technology available, they made close and careful observations and took meticulous notes. The community encouraged children to explore the world around them in as much detail as possible and post their observations and insights onto the Chirp-Net. In digital lingo, to enter an observation was to "chirp."

The Chirp-Net had created centers across the world where those who did not have access could upload their contributions through the written word or through voice, with visual and sound files. All entries could be read or heard in any language. To belong to the Chirp-Net meant chirping whenever you could, but you had to meet with a friend at least once a month for a "play date." Then, you explored the world together, chirping the most wonderful things you had seen recently. To play was to love. Observations were endorsed, often replicated, adding to their veracity. The chirps were detailed, playful, and always loving.

Chirp-Net was designed by Dhakiya, the Kenyan whiz kid who made a splash with her digital prowess. Mika of the Sioux Nation in North America had created an international language translator—so all children could write or speak in any language they knew and could read or listen to the messages in their language of choice. Gilma from Brazil extended the translation of common names into their scientific nomenclature. With the press of a button, you could find out if your observations were already in the public record. Indeed, so many new organisms, and more importantly observations of new organisms, emerged. Cyn from Greece helped catalog an extensive search engine. You could search with a word, sound, or picture. (Cyn is still working on a smell, taste, and touch digital translation!) Many groups formed through floral and faunal interests—Dandelion Dandies, Lily Lovers, Wobbly Warblers, Magnificent Monarchs.

Why did the powerful not just crash the server? Credit goes to the cynical acumen of some of the Chirp-Net's older members. The network of children had essentially shamed the powerful and older generations who had brought the planet to the brink but seemed unable to act. Did the young really have to save the old? What a damning picture, and what a profound responsibility the young were forced to shoulder. Shame seemed a powerful emotion. Second, the children and often grandchildren of the powerful were tenacious. They cajoled, convinced, and coerced. If this failed, they blackmailed. It worked, and the Chirp-Net flourished.

Noor, full after her meal, got on the Chirp-Net and carefully recorded her observations of the worms on the mulberry tree. Once she had identified the worms, she played the sound of their eating. She played it again and again. Her hunch was right—the worms sounded different. Was this a sign of alarm or innovation? Were the worms in trouble or had they learned to do something new? She wasn't sure. She uploaded the sound of her worms. Maybe the mulberry trees had changed? Her news would travel the waves of the Chirp-Net. Noor smiled. It was a good day. She signed off.

With the advent of climate change, the geographies of plants and animals were shifting. The plant detectives on Chirp-Net tracked the shifts and created detailed maps of the planet's changing biogeography. With a global volunteer detective force, children tracked migration patterns effortlessly. For them, this was not an experiment; it was survival. They were fighting for a future. Their efforts helped prepare for the changing climate—nurturing habitats for moving and migrating organisms. Front yards, local gardens, and open areas became the grounds for a dynamic experiment for global survival—a grand crowdsourcing for a public commons!

Revisiting the "Women in Science" Question: Diversity, Gender, and the Coloniality of Science

How might we bring the scholarship on gender and colonialism to bear on the "women in science" question? The impetus for bringing these fields into conversation with each other is the continued troubling and unyielding gap between projects and the literature on women in the sciences and the fields of feminist/postcolonial/decolonial analyses of the sciences. The literature on "women in the sciences" largely focuses on the underrepresentation of women through initiatives and projects, often around Women in Science and Engineering (WISE) programs at schools, colleges, and universities, as well as professional scientific societies. I use *women* because it is the term around which such politics are mobilized—both in the literature and in the attention of funding agencies. However, the arguments and goals are fundamentally about marginalized groups—those historically excluded from the hallways of science. The same arguments easily extend (and in fact do in many cases) to queer students, transgender students, and students with disabilities, as well as to students of color in the United States and for example Dalit and other marginalized caste groups in India. For feminists, it is critical to recognize that every category is itself diverse, to avoid having the most privileged of the group (straight, white, middle-class men in the United States and straight, Hindu, middle-class men in India) be the key beneficiaries.

Broadly, these initiatives have explored the reasons for the underrepresentation and devised strategies to counter the trend. Putting aside claims that women are biologically unfit to excel in the sciences, other explanations include the continued legacies of historical exclusions, lack of access to power and knowledge, lack of preparation, and lack of resources.[1] By now there is also a robust critique of the tired old metaphors explaining the attrition of women in the sciences as one of "leaky pipelines" and the solution as "plugging" those pipelines.[2] Similarly, many have critiqued diversity initiatives around gender and race in the academy.[3] They have ably demonstrated that much recent work on diversity has tended to "mirror" and reinforce the problems by not taking seriously questions of power and structures of racism, sexism, and casteism.[4] In Chandra Mohanty's astute summary, these efforts treat diversity as a "benign variation, which bypasses power as well as history to suggest a harmonious empty pluralism."[5] These histories are mirrored in the more recent fad, a compulsory mantra of "Diversity, Equity, Inclusion (DEI)" that has taken hold of universities and multiple public and private institutions in the United States. In short, the idea of "diversity" in its recent incarnations has been utterly domesticated and depoliticized. Everyone is for "diversity," everyone promotes "diversity," offering a lot of empty rhetoric and little fundamental change to the structures that prevent diversification.[6] Diversity is now largely a universal "good" that has lost its political roots in critiques of the social infrastructures of exclusion. To understand why this is so, we need to bring together the literatures on feminism, science, and colonialism.

WOMEN/GENDER IN/AND SCIENCE

How do we understand the gap between the framing of "women *in* the sciences" and "gender *and* the sciences"? The first deals with a focus on women scientists. Remedies include giving women greater access to resources, preparation, role models, mentoring, and networking opportunities. The second moves from the experiences of scientists to thinking about the gendered nature of science itself—that is, moving beyond questions of sex and gender to the infrastructures of science.

In *Ghost Stories for Darwin*, I lay out the key critiques of efforts for women in the sciences, summarized here.[7] In the United States, institutional efforts to diversify workplaces involve government agencies such as the National Science Foundation and the National Institutes of Health, as well as private organizations such as the Association of Women in Science, the American Association of University Women, the Women in Engineering Professional Action Network, the Society of Women Engineers, and women's task forces and caucuses within professional societies of scientific disciplines. All these institutions and groups have funded projects and programs designed to increase the recruitment and retention of girls and women in the sciences. More recently the work has fallen under the rubric of Diversity Equity Inclusion efforts at universities and organizations.

While these efforts have ushered in an increase of marginalized groups in the academy (although nowhere close to parity), accounts by marginalized groups chronicle the deep-seated sexism, racism, homophobia, classism, ableism, and casteism that continue to pervade the academy. At the heart of the critique in feminist STS is that initiatives for women in the sciences have been grounded in and reduced to a narrow, liberal idea of equity or equality of representation as the primary intellectual, political, and ideological strategy.[8] This comes from a larger cultural adoption of a principle of equality of *opportunity* rather than equality of *outcome*—the latter being the more radical commitment to social justice. It is well worth pondering why demographic numbers are the cornerstone of focus rather than cultural or structural transformation. Despite equity feminism, parity across the sciences remains a distant goal. The professoriate remains male dominated, especially at prestigious institutions and in senior ranks. Why?

The problem, as the early scholars of feminist STS pointed out, is an exclusive focus on women instead of on the gendered and racialized nature of the fabric of science itself. The exclusive focus on increasing the number of women has led to a myopic and singular focus on equity and parity and a narrow vision of feminism. The "women in science" discourse has been haunted by a path of "relentless linear progressivism" as the result of particular discursive formations.[9] There have been some changes in the predominant language in the past three decades, but interrogating the underlying assumptions about women is rare. The shift from exclusion to inclusion is marked by a shift from exclusion based on claims of the innate, biologically based inferiority of women's scientific abilities to a politics of inclusion dominated by policies that address women's biological bodies and gendered roles as wives, daughters, and mothers.

For example, let us consider the push for "female-friendly" policies. Despite their progressive ambitions, in emphasizing issues of reproduction and family, advocates of "family-friendly" policies reassert women's reproductive potential as a central concern, marking "female difference" as hypervisible while leaving the worlds of masculine epistemic cultures untouched. Why does there continue to be an intense focus on increased productivity and an unhealthy workaholic culture? Why are these norms rewarded? Why does excessive competition continue to haunt academic culture? To be sure, it is neither desirable nor persuasive to articulate an "anti-family-friendly" perspective. But a strategy that consistently emphasizes women's role in the family reinforces essentialist ideas about women. After all, we could promote the need for balanced lives, be they about family or other pursuits and hobbies. By refusing to challenge the normative model of the male as the ideal scientist, which insists on a productivity that can be achieved only through very long hours, a singular dedication to work, and an exclusive focus on one's profession, the core of science as a pursuit by those who are unencumbered by other responsibilities persists. David Noble aptly describes scientific culture as a "world without women." As he argues, with its roots in a Christian clerical tradition, science emerges from an "exclusively male—and celibate, homosocial, and misogynous—culture, all the more so because a great many of its early practitioners belonged to the ascetic mendicant orders."[10] In short, science has traded the clerical robe for the lab coat.[11] In this context, we can see why solutions such as mentoring women to negotiate the normative model, or rewarding women who accept the normative model, or retaining women through "special accommodations" that increase their workplace flexibility (part-time appointments with administrative or teaching responsibilities) or automatic pregnancy and family leave are inadequate to deal with this deep historical legacy. The original standards for excellence are never challenged; the solution is about helping women conform to them. In extolling feminine virtues, the "women in science" discourse endlessly reinscribes women in relation to femininity and the domestic reproductive sphere.[12] The goal for feminist STS is not just parity and equity but a transformation of science into an enterprise that is inclusive not only of people but also of ideas, of multiple scientific epistemologies, methodologies, and methods.

COLONIALITY OF POWER

If gender is a construct that shapes and has been shaped by powerful institutions like science, where does gender come from? Here, a robust scholarship on the coloniality of power and gender is useful. Fundamentally, the concept of coloniality helps us understand how colonialism shaped institutions of knowledge and ushered in a new world order. Feminists argue that coloniality also ushered in a coloniality of gender, or the "modern colonial gender system."[13] Western conceptions of gender roles were imposed across colonial worlds. These gender roles continue to shape scientific demography; the clerical robe continues to shape the lab coat. How and why does the concept of the coloniality of power and gender help us understand the coloniality of science, and especially the "women in science" question? I should make clear that much of this work is more general and not focused on botany. I want to highlight five major themes.

First, it helps us understand why biology and the body remain so entrenched in our vocabularies. Ideas of sex and gender as inviolable bodily features seem resolutely fixed. Colonial biopolitics transposed a western conception of sex, gender, and race on the body—theories of reproduction, anatomy, physiology, and capacities for rational thought and behavior. The discourse on women in the sciences is deeply embedded in these histories. Even when projects work to dismantle entrenched structures, efforts largely return to make room for the feminine reproducing body in science. Understanding the coloniality of power and gender, and their role in the formation of modern science, forces us to ask different sets of questions about dismantling the very rubric of sex, gender, and race. So, for example, instead of asking how we might accommodate the reproducing body in order to fulfill expectations of the ideal sciences, what if we questioned those very expectations? What if the ideal scientist were not expected to work unhealthy hours? What if we rewarded scientists who lived full lives, who spent time connecting with their communities, and who were good mentors? What if the workaholics were not heroes but pathological characters? What if we did not reward the overproduction of (often mediocre) publications as a mark of excellence? What if we dialed down expectations of productivity? There are so many ways we might reenvision what it means to be a good scientist. Science is a complex, communal and intensely social enterprise, yet most of the laboring bodies remain anonymous while the laboratory heads and senior scientists receive all the glory. Like the white explorers who claimed discovery of plants and animals long known to colonized civilizations, modern science privileges the heads of labs and claims for them the labor of lab workers. Data shows that western scientists continue to "parachute" into formerly colonized countries to collect samples and then return home to process the specimens and publish. Thinking about the coloniality of science helps us understand science's ongoing feudal and colonial frameworks. If botanists are serious about confronting colonial histories, scientists in the global north could collaborate with those in the global south, copublishing and helping build infrastructures and train students. Like colonial extraction, modern botany continues to act like a "vampire" as it sucks resources from colonized nations.[14]

Second, the coloniality of science reveals its legacies in the uneven geography and demography of science. The vast majority of contemporary science that receives acclaim remains in the west. Nobel laureates, even those born in formerly colonized nations, work in the west. Two sets of issues shape this demography. The rules of what is "good" science and what science has impact are still made in the west. Decolonizing science means that we need to rethink the rules of the game. The infrastructures of science are also deeply unequal; global inequality within and between countries is striking. Global financial regimes dominated by the west shape some of this inequality.

In a postcolonial nation like India, which just celebrated its seventy-fifth year of independence, local elites must share in the blame. So much of Indian science and its education system remain remarkably colonial, centering the west and its canon. As I described earlier, a scientific education is often a lesson in alienation for Indian students. Even as an English-speaking student, I felt this alienation. Poems described plants and animals I had never seen, and examples in my science classes presented knowledge as fact rather than discoverable. The system pushes the student to study by rote rather than experience. Science needs to be connected to student lives if it is to be meaningful. C. V. Raman puts this well: "I feel it is unnatural and immoral to try to teach science to children in a foreign language. They will know facts but will miss the spirit. People forget that science is culture, a child must learn from science many things; a true scientific attitude not only engenders objectivity but also spiritual values. We MUST teach science in the mother tongue. Otherwise, science will become a highbrow activity. It will not be an activity in which all people can participate. It will not percolate into the heart or blood of our people."[15] Circumventing the fraught politics of language in India, I take from Raman the important idea that science must be taught so it can be claimed by a student as her own, a knowledge system that has the potential to evoke curiosity, play, application, and innovation.

Third, the coloniality of science gives us a glimpse of colonial hubris. Colonists felt empowered, in fact divinely anointed to enslave, conquer, exploit, and colonize labor and land. They extracted the rich resources of colonies such as spices and herbs. They claimed local knowledge and renamed plants after white scientists, explorers, and rulers. They built plantations and enslaved and transported peoples. They decimated local knowledge systems, practices, and cultures to install a colonial, European system as *universal science*. Despite its many appropriations of local knowledge systems, it claimed them under the mantle of "western science" with no acknowledgment.[16] What we understand as modern science today—its philosophies, epistemologies, and knowledges—developed in the aid of colonialism, in attempts to better explore and exploit the resources of the colonies.[17] In the history of science, botany continues to be presented as a western invention, a civilizing and modernizing force. Even in postcolonial nations, and certainly in my education in India, science and technology was imagined as the engine for Indian modernity. I studied mostly about western scientists, rarely any Indian ones. Yet historians have demonstrated that during colonialism, natives resisted, and knowledge circulated within and between empires.[18] I did not learn about any of this in my Indian curricula, only a parade of scientists who were almost all white men. Science is not singularly western but is embrangled with the global.[19] If we are serious about decolonizing science, we must decolonize the history of science. We need to center the new historiographies of science from postcolonial, indigenous, and queer historians, a renarration of the history of science that eschews any neat divisions of east and west.

Fourth, the hubris of western science endures, with profound consequences for women scientists in formerly colonized worlds. Consider "parachute" science. A recent study analyzing over three hundred thousand papers published over the last two decades in the plant science literature concludes,

> Our analyses reveal striking geographical biases—affluent nations dominate the publishing landscape and vast areas of the globe have virtually no footprint in the literature. Authors in Northern America are cited nearly twice as many times as authors based in Sub-Saharan Africa and Latin America, despite publishing in journals with similar impact factors. Gender imbalances are similarly stark and show remarkably little

improvement over time. Some of the most affluent nations have extremely male biased publication records, despite supposed improvements in gender equality. . . . Taken together, our analyses reveal a problematic system of publication, with persistent imbalances that poorly capture the global wealth of scientific knowledge and biological diversity.[20]

How is this description of science today any different from colonial-era practices? Even the solutions reinscribe the same colonial histories. For example, well-meaning projects that bring third world scientists into hallowed spaces of US science end up not as opportunities for collaboration but what Carol Ibe calls "the tours of the impossible."[21] What can a few days, weeks, or months of spectatorship of western sciences offer scientists in countries with impoverished scientific and publishing infrastructures? It is worth reconsidering such tours, which are often seen as a progressive practice and are increasingly a cornerstone of many scientific conferences.[22]

Similarly, take the example of herbaria that house botanical specimens or the original "type" that is required for classification. As many as 96 percent of certain families are housed in the west (see chapter 4).[23] Yet the west remains inaccessible to many because of differing resources. The same can be said of museums and their acquisitions. Decolonizing science means paying serious attention to questions of reparations and the rematriation of resources.

Fifth, there were and are rich and vibrant knowledge systems of what we might call mathematics, science, and medicine in the precolonial and the contemporary world.[24] Yet these systems are minimized and relegated to the world of ethnosciences. So much of indigenous and local knowledges resides in the hands of marginalized peoples, especially women, who still live close to the land. The coloniality of science challenges us to understand western science as a European ethnoscience. Decolonizing science compels us to reckon with the lost histories of world knowledge systems. As we have seen, centering the histories of gender, race, and caste is critical. We need more robust and systematic tools to compare and contrast incommensurable knowledge systems. We must imbue our students with the critical thinking tools to ascertain the veracity of knowledge. Simply falling back on "science" as the only arbiter of truth is too facile a response. We need critical thinking skills to evaluate all knowledge claims.

For example, I am deeply troubled by contemporary efforts in India to claim Vedic sciences as modern science in the name of anticolonial politics. Stories of Ganesha, the elephant-headed god, provided as "proof" that plastic surgery existed in mythological times strain credulity. Mythological tales are important in their allegorical imaginations, and we must value them for that. Collapsing them into science does damage to both science and mythology. The rich traditions of science in South Asia (such as astronomy, mathematics, medicine, and philosophy) are rarely included in school textbooks. Any project of decolonization must celebrate the multireligious histories of India's scientific heritage. Our education systems have a long way to go to inculcate both curiosity and critical thinking tools for our times.

Maori intellectual Linda Tuhiwai Smith defines decolonization as "a long-term process involving the bureaucratic, cultural linguistic and psychological divesting of colonial power."[25] We need a renewed politics to undertake this task. We need more robust theoretical and interdisciplinary tools, epistemologies, methodologies, and methods to both disentangle the colonial myths that masquerade as histories and reentangle a history of the complex circuits of global colonialisms. We must reimagine knowledge systems that remain haunted by the legacies of empire but that have broken the shackles of empire to dream new liberatory worlds.

PART TWO

Kinship Dreams

CLASSIFYING PLANT SYSTEMATICS

Give name to the nameless so it can be thought.
The farthest horizons of our hopes and fears are cobbled
by our poems, carved from the rock experiences of our daily lives.
AUDRE LORDE | *Sister Outsider: Essays and Speeches*

CHAPTER THREE

The Categorical Impurative

Names, Norms, Normings

Do you know that even when you look at a tree and say,
"That is an oak tree," or "that is a banyan tree," the naming of the tree,
which is botanical knowledge, has so conditioned your mind that
the word comes between you and actually seeing the tree?
To come in contact with the tree you have to put
your hand on it and the word will not help you to touch it.

JIDDU KRISHNAMURTI | *Freedom from the Known*

Purism is a de-collectivizing, de-mobilizing,
paradoxical politics of despair. This world deserves better.

ALEXIS SHOTWELL | *Against Purity*

Prologue

As a South Asian immigrant in the United States, I often meet my second-generation counterparts. When I first came to this country and asked their names, I was amused, and sometimes bristled, when they slightly or completely "mispronounced" their names. While I thankfully suppressed my impulse to "correct" them, it was only well into my forays into ethnic studies that I came to question the idea of "mispronunciation." I had to ask myself, Who made me the judge? As I thought about my travels around India, I remembered how the same name is pronounced differently in different regions of the subcontinent. Indeed, on many an occasion, people from various states from northern India corrected me when I proclaimed my name, written and spoken, as Banu instead of Bhanu. This is a Sanskrit name, they told me. Maybe so, I learned to respond, but I'm Tamilian. Tamil does not have the letter or sound "bha." This is how my parents spelled my name on my birth certificate. I'm claiming it! With the rising Hindi chauvinism in India, I've dug even deeper in claiming the Banu—it may be a Sanskrit word, but this is how I pronounce it. Yet many (including some in my family) still insist on writing to me as and calling me Bhanu. Names are personal, I have come to recognize. I ought to pronounce a name as given and requested by the bearer of the name, not through sanctified politics of purity. The "correct" name is always shrouded in structures of power—questions about authenticity, about right and might. As Max Weinreich quips, "a language is a dialect with an army and navy."[1] More importantly, in formerly colonized countries like India, where English remains one of the official languages, local languages are transliterated into the English alphabet. You could be Rina or Reena, Sita or Sitha, or Seeta or Seetha. To complicate matters, with the rise of Hindu nationalism, India in recent years has seen a huge uptick in numerology, leading to exuberant additions and omissions of vowels and consonants to take numerological considerations into account. Just consider Bollywood actors: Devgan became Devgn, Yash became Yassh, Khurana became Khurrana, Irfan became Irrfan, and Sanjana became Sanjjanaa.[2] With nationalist fervor, cities have embraced renaming. Did I grow up in Calcutta, Bombay, Madras, and Calicut or in Kolkata, Mumbai, Chennai, and Kozhikode?

My experiences in India and then an immigrant in the west are shaped by histories of hybridities. I have developed a flexibility and openness to names and naming and have come to terms with the potent politics behind multiple spellings, naming, renaming, and ever-changing pronunciations. Naming is a political act, and renaming likewise is not only symbolic but a kind of reckoning with the past. I'm also rather unmoved by claims of the sanctity of origin stories, of properness, of the immutability of tradition or culture. To me, words evolve, contexts change, and the world moves on. Today when I meet a second-generation Indian immigrant, I take their name as a reflection of their history and not a travesty of language.

Bringing Order to Plant Worlds

My experiences with names and naming shape my analyses and reflections on botany. Botanical naming and nomenclature are deeply intertwined with how plants are organized into classificatory systems (fields of taxonomy and systematics). While in the sciences, papers published five years ago are considered "old," nomenclature is one subfield that remains rooted in history.[3] Naming and taxonomy have ongoing and interrelated histories, and I explore them both in this chapter.

Most of the vast literature in the field of taxonomy describes the goal as bringing "order" to the diversity of nature. I've been struck by how often I've seen some version of the following sentiments:

> Man is by nature a classifying animal. His continued existence depends on his ability to recognize similarities and differences between objects and events in his physical universe and to make known these similarities and differences linguistically. Indeed, the very development of the human mind seems to have been closely related to the perception of discontinuities in nature.[4]

As the story goes, because we are a classifying species, people across the globe brought "order" to the natural world by naming local plants and animals. But what is order? A group that categorizes the world into three categories—edible, inedible, and poisonous—has introduced an important practical and life-sustaining order. There is order everywhere, but the "order" brought about by botany is different and is deeply implicated in the histories of colonialism. As western explorers traversed the globe, they encountered diverse sets of nomenclatures and "order." Prescientific societies (including in the west) also classified, named, and identified living things. In botanical lingo, these are labeled "folk taxonomies," a patronizing and marginalizing term.[5] The trouble with folk taxonomies, the argument goes, is that "these names vary in a chaotic fashion from place to place and time to time." So early in Western history arose "a professional group whose duty it was to bring order to this chaos."[6] This group, systematists or taxonomists, brought botanical order. Their subdiscipline attempted to classify all living organisms into a universal, single, and stable system. "Lumpers" lump organisms by similarities and "splitters" note differences as the grounds for classification. Tensions remain between lumpers, who define taxonomic groups broadly (fewer groups but larger numbers within each), and splitters, who multiply the numbers of groups and names (many groups with small and precise composition).[7]

Why do we need names, nomenclatures, and classifications? It is useful to begin with some definitions. Scientific names are based on a Linnaean binomial system. Plant naming and nomenclatures are regulated by a community of taxonomists—particularly those serving on the committee to create the *International Code of Nomenclature for Algae, Fungi, and Plants.* The members, mostly western scientists, meet regularly and set guidelines and rules for an active and ongoing process of nomenclature. Taxonomy is the theory and practice of identifying, describing, naming, and classifying organisms—that is, identification of taxa, or scientifically codified groups. While the terms *taxonomy* and *systematics* are often used interchangeably, Ernst Mayr defines *taxonomy* as "the theory and practice of classifying" and *systematics* as "the study of the diversity of organisms," grounded in describing evolutionary relationships.[8] Both taxonomy and systematics use morphological, behavioral, genetic, and biochemical observations to classify organisms.[9]

Why do we need a singular universal system? And if we do need one, why did the colonial west get to determine it? Why is the scientific west never a "folk" language? The very language I'm forced to use belies the issue. What is the problem if people around the world develop their own locally coherent folk taxonomies *as* science? Why can't we have a diversity of botanical languages, like we have a diversity of linguistic ones? Who needs universal stories? Is the scientific story truly a universal and coherent one? Is it "superior" to folk taxonomies? If so, how and for whom? Who gets to tell the scientific story? In what language is it told? Is there consensus about this story and its attendant systems? As you will guess, at the heart of the story are colonialism and coloniality—colonial exploration, colonial expansion, colonial profits, and colonial rule—all of which shaped the field of botany and the modern plant sciences, especially the naming, nomenclature, and classification of plants. Even in postcolonial nations (as we see in the next chapter), botanical infrastructures remain mired in western and colonial logics.

A Brief History of Plant Nomenclature, Taxonomy, and Systematics

I begin with a brief history to show how colonial ideas shaped modern botany. This chapter is organized more or less chronologically: "Folk Taxonomies and the Prehistory of Plant Naming," "Linnaeus and the Legacy of Binomial Nomenclature," and "The Rise of Evolutionary Sciences and the Advent of DNA." In each section, I first narrate the history of the field and then reflect on its colonial afterlives. I follow with scientific infrastructures of naming and their salience and shortcomings. Finally, I conclude with a discussion of the "tree of life." What began as a study of the morphology of plants has increasingly become a study of the molecular. I end with a discussion of some key debates and themes in the history of plant taxonomy and reflections about botanical decolonizing.

FOLK TAXONOMIES AND THE PREHISTORY OF PLANT NAMING

The history of botany begins with early and diverse understandings of plants across the world thousands of years ago (deemed folk knowledges). The roots of scientific botany, however, are usually traced to the Greeks and Romans. Genealogies acknowledge a long list of individuals who attempted to order plant worlds; they are exclusively men and from the west. Once plants were named, scholars attempted to create a taxonomy or system to organize them, looking for similarities and differences. Some of the key figures include Theophrastus (370–285 BC), often referred to as the "father" of botany, who classified plants by their form, as herbs, shrubs, trees, and so on; Andrea Caesalpino (1519–1603); Gaspard Bauhin (1560–1624); John Ray (1628–1705); and Joseph Pitton de Tournefort (1656–1708).[10] The early history of botany was attached to medicine, focused on medicinal herbs. Indeed, most great botanists from the sixteenth to eighteenth centuries, including Linnaeus, were professors of medicine or practicing physicians.[11] Early botany in the sixteenth century linked botanical medical treatments with plant habitats. For example, God put willows in wet habitats because men in wet habitats, prone to rheumatism, could be served by the willow.[12] A closer look at the scientific history of taxonomy therefore betrays its own western "folk" roots since so much of plant understanding in the natural world relied on anthropocentric elements.[13]

Before the seventeenth century, homological thinking (prevalent during the Renaissance) was shaped by observations of correspondences and similarities among living things.[14] Botanical narratives reveled in the deep and rich connections among life on earth. Yet pre-Linnaean and pretheoretical taxonomies did not reduce nature into one "system."[15] Rather, these were embrangled worlds structured by a vertical and hierarchical principle, where "society had a 'natural' order and nature had a 'social' order."[16] For example, the *Doctrine of Signatures* attempted, prior to the end of the sixteenth century, to predict effective herbal remedies for various human illnesses.[17]

Michel Foucault traces the epistemological shifts from this earlier order into one marked by rational taxonomies and careful attention to the kind and categories of organisms—human, animal, plant—and eventually divisions between them. Foucault argues that this modern epistemology separated word and language from signs and forms of truth, thus dividing history from science.[18] While "premoderns" found endless, rich, and enchanted connections between nature and culture, modernity ushered in an absolute dichotomy between nature and culture.[19]

Folk taxonomies served botanists well until the great Age of Exploration (a euphemism for colonialism if I ever heard one!). Plant hunting in the late eighteenth century and most of the nineteenth unleashed a flood of "new" species that Europeans had never encountered before. For hundreds of years, Endersby argues, European naturalists believed that the lists of plants and animals they had inherited, especially from the Greeks, were largely complete.[20] When colonists discovered a treasure trove of new plants, animals, and peoples that they had no names for, they gave them names. At first Europeans gave these organisms new names, often wildly inaccurate translations and transcriptions of indigenous names or very long descriptive names. Many amusing stories are chronicled in the literature. For example, Dutch explorers of the Americas listed a long Arawak plant name that later turned out to mean, "I don't know this one so I'll have to ask my uncle."[21] Or take the Latin name for one particular plant species: *Acaciae quadammodo accedens, Myrobalano chebulo Veslingii similis arbor Americana spinosa, foliis ceratoniae in pediculo geminatis, siliquoa bivalve compressa corniculate seu cochlearum vel arietino-*

rum cornuum in modum incurvate, sive Unguis cati ("A spiny American tree, in some ways resembling Acacia, similar to Vesling's *Myrobalanus chebulae*, with *Ceratonia* leaves in pairs at the pedicle, a silique with two valves, which is compressed, and horn-shaped or curved like the horns of snailshells or rams' horns, or like a cat's claws").[22] As Tod Stuessy argues, "Just about everything under the Sun was used."[23] These is why the simplicity of Linnaeus's system emerged as so useful.

The colonists rapidly named their "discoveries" in the colonized worlds, creating a virtual "new Babel."[24] But the problem for colonists was that the same plant or animal had different names in different parts of the colonies. How to tell if the aromatic spice in one region was the same as the one in another? By the eighteenth century, the colonists felt the urgent need for a consistent system. After all, how could they exploit natural resources efficiently if they did not know what plants existed where? Might they be able to grow these in their homelands? This created the grounds for a botanical nomenclature system with standardized names, well-developed life history details, good ecological plant characterizations, and a standardized hierarchical system of organization. Eventually a scientific infrastructure of botanical nomenclature emerged—standardized diagrams, herbarium sheets that displayed the plant with strict protocols, and key biological and contextual knowledge about the plant. The herbarium sheets traveled to find homes in colonial institutions, where they often remain to this day.

Much of colonial exploration was about utility. Ernst Mayr argues that when botany moved from its roots in medicine to become an independent science, and with the great colonial expeditions and the burgeoning of new species, the first concern was bringing order; eighteenth- and nineteenth-century botany was largely taxonomy.[25] Early taxonomy focused on classification through morphological similarities and differences. This transformed into plant systematics with its focus on evolutionary relationships, the topic of the next section.[26] Later, molecular technologies further transformed the field.

COLONIAL ROOTS AND AFTERLIVES Several aspects are worth noting in botany's early historical narration. First, just as Columbus "discovered" America, so did white colonists "discover" plant species in the colonies. Species widely used for centuries by local peoples emerge in western narratives as ones of "discovery," and the discoverer's role is forever enshrined in the scientific names of the species.

The vast apparatus of knowledge systems of indigenous medicine become fodder for western innovation. Hypocrisies abound. Indigenous systems and peoples are deemed prescientific, barbaric, and lacking civilizational value, yet their vast knowledge systems and natural resources were (and are) extracted by colonists.[27] Colonial bioprospecting is today's biopiracy.[28] A western-dominated botanical establishment and infrastructure continues to set the terms of the debate.

Colonial logics enshrined plant official names in Latin (or latinized terms). With the rise of climate change and the continued onslaught against forests of the world, there is an obsessive anxiety about the loss of unnamed species. In 2015 alone, about 2,034 new species were discovered.[29] Some botanists estimate that of over 8.7 million terrestrial species, only about 14 percent are presently known to science (not counting the diverse world of protists, fungi, bacteria, and viruses).[30] In short, we need to speed up our taxonomic record and the "discovery" of new species to better chronicle and mourn the species we are losing![31] A profound irony. This obsession with counting species (producing data) betrays an enduring logic of coloniality: a study of plants without people.

LINNAEUS AND THE LEGACY OF BINOMIAL NOMENCLATURE

In the history of botany, and indeed biology, the name of Linnaeus looms large. Undoubtedly one of the most celebrated and influential figures, Carolus Linnaeus (1707-1778) is regarded by disciplinary historians as the "father" of systematics. Linnaeus is important for two reasons—he created the first system of binomial nomenclature of plants that endures to this day, and he produced a classification and organizational system for plants. He also created a hierarchical model that followed Aristotelian representations of nature in three kingdoms—minerals, plants, and animals. In 1735, he published the *Systema Naturae*, where he organized biological life into a hierarchical set of groups—species into genera, genera into orders, orders into classes. His system was simple, a sexual system of classification based on the number and position of the reproductive organs in a flower.[32] Linnaeus believed *Nomina si nescis, perit et cognition rerum*: "If you know not the names of things, the knowledge of the things themselves perishes."[33]

Historians regard Linnaeus's greatest contribution as his method of naming species. This method, called binomial nomenclature, gives each species a unique, two-word name consisting of the genus and species name. For example, humans are *Homo sapiens*, or "wise human," a name that references our big brains.[34] Linnaeus classified 7300 species into 1098 genera.[35] Many local names gave way to a binomial latinized official name. He named his favorite flower after himself, *Linnaea borealis*. Linnaeus named so many plants and animals that he has been called "God's Registrar."[36] It is poetic justice that *Linnaeus* is a latinized form of *Linné*.

Endersby argues that the simplicity of the Linnaean sexual system helped it become popular, influential, and critical to the continued expansion of European knowledge of the world.[37] But his sexual system came at a price. Because of Linnaeus's focus on the sexual anatomy of flowers, botanists such as Johann Georg Siegesbeck accused him of subjecting his students to immoral influences.[38] Other biologists believed that taxonomy ought to reveal more about the world than merely create distinctions.[39] Some challenged the curious and singular focus on the reproductive system; Linnaeus paid no heed to the many other plant features that could produce a more "natural" system. At one point, some scientists complained that despite decades of work, "nature's order had still failed to reveal itself."[40] Even for Linnaeus, such a revelation was the ultimate goal. In 1751 in the *Philosophia Botanica* he wrote: "The true beginning and end of botany is the natural system," a system in which "all plants exhibit mutual affinities, as territories on a geographical map."[41]

The goal of the natural system was to pay attention to the overall morphology of a plant, not just its flowers or fruits. The British generally remained enthusiastic users of the Linnaean system, making it popular in Britain and its colonies, histories particularly influential for botany. However, tensions remained between several classificatory systems. Today, the rules by which we name plants still depend largely on Linnaeus, although for plant families, A. L. de Jussieu's classification in his *Genera Plantarum* of 1789 remains important.[42] By 1799, there were fifty-two different systems of botany, each taxonomy emphasizing a different part of a plant or plants' geography.[43] Ultimately, by the end of the eighteenth century, the Linnaean system beat out the others and emerged as the norm.[44] Despite the advent of evolutionary methods and DNA technologies, and despite its many critics and shortcomings, Linnaean classification remains remarkably resilient.

Empires, especially the Spanish, embraced the Linnaean system. Further, as King astutely argues, it seemed functional precisely because it relied on a social understanding of marriage and normative modes of sexuality.[45] Linnaeus's analogies between plants and humans, particularly about sexuality and the distinctions between male and female, made his work accessible.[46] He devised his *methodus propria* (proper method) as a temporary solution.[47] Only later in life did he realize that this temporary shorthand of a system would become his enduring legacy.[48] Linnaeus explicitly included rather than excluded women from the study of botany.[49] He reached out to lay and learned audiences, encouraged translations of his work, and in short, worked hard to democratize the accessibility and usefulness of his classification.[50] This broad dissemination and popularization were immensely successful. Unlike other classificatory systems, Linnaeus's caught on, creating the emergence of a "botanical vernacular" in literature and popular culture—perhaps the combined result of its relative simplicity as a taxonomical system and its vivid imagery of the sexual.[51]

The power of the Linnaean system is that it is both classificatory and nomenclatural.[52] Linnaeus published his classification system in the 1700s. Since then, many new species have been discovered. Initially, Linnaean ranks (kingdom, phylum or division, and so on) were thought to be objective because to him they were the plan of the creator, but with time, the Linnaean system was seen as not natural because it focused only on the sexual parts of plants, and thus needed revision.[53] Linnaeus's success was never absolute—for example, the Jussieuean system found favor among professional proponents in the 1830s—but Linnaeus endured in the "botanical vernacular." Thus, King argues that while the line between professional and amateur natural history hardened in the first half of the nineteenth century, Linnaeus retained a strong presence in the vernacular but not the professional realm of botany.[54]

COLONIAL ROOTS AND AFTERLIVES The power of naming in Christian cultures is a divine power. In Genesis, Adam is given the power to name the animals. In the botanical realm, botanists claimed a "natural" and God-given power to name.[55] This power emerges in the language and practices of imperialism. Christopher Columbus named the first American island he encountered "*San Salvador*, in remembrance of the Divine Majesty, Who marvelously bestowed all this."[56] Not only did Columbus take possession on behalf of God and the Spanish monarchy, but he claimed a righteous power to do it even while knowing that the local Indians called the island "guanahani." In asserting his power, he erased native nomenclature.[57] This colonial practice of naming is at the heart of scientific epistemology—naming places a barrier between observer and observed, a separation between scientist and nature. Here, the scientist is all supreme.[58] Latinized scientific names, unlike local names based on plant characteristics, render plants alien. This attitude is pervasive; organisms were named by the colonizers to celebrate European legacies—God, Queen, King, plant "discoverer," friend. Indeed, the very term *animal* (or, for that matter, *plant*), bundling together disparate creatures despite their many differences, speaks to this power.[59] As Prosek reflects, "I think what bothered me was his hubris. . . . Plants and animals already had names in indigenous languages, and Linnaeus, in a show of imperialism, renamed them with his Latinisms. He believed he could take nature—holistic, fluid and constantly changing—and fragment, label and systematize it."[60]

Linnaeus was a superb promoter of his work. Indeed, the global exchange of seeds and plants created a vibrant political economy for Linnaean botany.[61] He was deeply engaged in global botanical exchanges and "trade relations," establishing himself as "the most consummate botanist."[62] Lisbeth Kroerner's excellent biography of Linnaeus describes his expedition to Lapland, the northernmost region in Sweden, between May and October in 1732 as an important part of his scientific career, especially in how he positions himself vis-à-vis the indigenous.[63] By the 1600s, a century before Linnaeus ventured forth, the indigenous Sami had already lost their native religion and been converted to Christianity by missionaries; they had become "tourist guides of their own culture's destruction."[64] Given this, it is particularly ironic how Linnaeus exaggerates the hardships he encountered and overestimates the miles he traveled (he was being paid by the mile). The Sami emerge in Linnaeus's *Flora Lapponica* as primitive and untouched, and they and their lands are portrayed as alien and strange. Linnaeus writes: "It seemed as if I were entering a new world; and as I climbed higher I did not know whether I was in Asia or Africa, for the soil, the situation and all the plants were strange to me. . . . There were so many that I feared I was taking away more than I would be able to deal with."[65]

His description and claims about the indigenous Sami are particularly illustrative. Linnaeus presented his Lapland travels as "an empirical enterprise and an adventure upon a *tabula rasa*." He dismissed older texts and remade the Sami as savage modeled on "international ethnography on indigenes of America, India, and Western Africa as well as on Ossentians, Samoyeds, Inuits and Kalyks." Yet he spent only eighteen days of his five-month journey outside the homes of Swedish homesteaders or on the coast. Koerner notes,

> Linnaeus mythologized his sub-Artic travels into a formative encounter with an Edenic "wild nation" (Sami reindeer herders) and a cross-cultural encounter between "high" and "folk" science. He then used this performative narrative to enter Dutch learned circles. Yet the Lapland journey was also part of the colonization of "our West Indies," as Scandinavians termed their Artic frontier. And this colonial venture in turn was predicated on erasing indigenous culture, as the "wild" Sami and their herds were chained to the engines of cameralist industry.[66]

Geoff Bil extends this insight to argue that the act of plant classification was profoundly a naturecultural project. European fieldworkers saw local plant names as a potentially "lucrative storehouse of indigenous knowledge."[67] Beyond botanical fieldwork, indigenous plant names opened a window into producing colonial ideas of indigenous cultural difference. Bil chronicles the rich and complex histories of plant biology and ethnology where indigenous phytonyms become the basis of claims of racial and cognitive differences of local peoples. For example, Alexander Humboldt claimed an inverse correlation between plant diversity and cultural sophistication. European societies, he preposterously argued, compensated for their impoverished botanical heritage with richer languages and intellects. Thus, Linnaean plant names helped produce stable, mobile, and universal systems, critical to the enterprise of mapping "imagined and actual worlds" of plants and people.

The Eurocentric and colonial disposition of Linnaeus endure. The nomenclature adopted by the Vienna 1905 International Congress of Botany echoes the colonial legacies of Linnaeus. Consider the following:

V. Botanists who are publishing generic names show judgement and taste by attending to the following recommendations: . . .

   c) Not to dedicate genera to persons who are in all respects strangers to botany, or at least to natural science, nor to persons quite unknown.

   d) Not to take names from barbarous tongues, unless those names are frequently quoted in books of travel, and have an agreeable form that is readily adapted to the Latin tongue and to the tongues of civilized countries.[68]

While colonial ambitions were many, colonialism was often rationalized as a "civilizational" mission of lesser humans.[69] The above criteria—with terms such as *judgement, taste, persons strange to botany, barbarous tongues*, and *civilized countries*—are barely disguised code for colonially professed "superior" European norms distinct from those of indigenous and colonized peoples. By this logic, the barbarous, non-Latin-tongued people could be lovers of plants, in their primitive ways, but not the purveyors of botany.

THE RISE OF EVOLUTIONARY SCIENCES AND THE ADVENT OF DNA

Today, the impulse to name organisms or create more accurate "natural" classificatory systems of plants is no longer a task to uncover divine order (as for Linnaeus) or for the ease of living. Rather, the overarching goal of systematic biology is to produce a comprehensive and accessible catalog of life on Earth.[70] While traditional taxonomy encompasses the description, identification, nomenclature, and classification of organisms, the subfield of systematics is dedicated to evolutionary relationships between organisms. The rise of evolutionary sciences, and the advent of modern systematics, reframed biological classification as the evolutionary relationships of taxa. Phylogenetic descent or evolutionary classification became the grounds for taxonomy, and genetic information replaced morphology as the critical site of data. The search for a "natural system" is ongoing.

With technological advances in the 1950s, molecular biologists found confidence bordering on imperial zeal, creating a rift with evolutionary biology.[71] Evolutionary biologists believed that natural selection worked at the level of the organism, while molecular biologists insisted that genetic data were key. Molecular biologists argued that mutations in the genetic code were the key source of variation; evolutionary biologists, on the other hand, argued that natural selection, not DNA, acted primarily on the organism. Scientists in the fields upheld the superiority of their chosen field, while some tried to find a compromise where the two fields would complement each other.[72] In a wonderful summary, two molecular biologists argued that DNA is not a passive carrier of the evolutionary message: "Evolutionary change is not imposed upon DNA from without; it arises from within. Natural selection is the editor, rather than the composer, of the genetic message."[73]

In the mid-1980s computational power transformed biology into a data-driven science. The wet labs (full of lab benches) were transformed into dry labs (full of computers).[74] While wet labs and their work continue, hardware and software are critical in an increasingly data-driven field. Systematics, a field founded on the principles of evolution and a goal to "discover" the most accurate phylogeny of life on earth, moved to using vast data sets to ascertain evolutionary relationships.[75] It is assumed that the more characteristics one includes, the greater the accuracy. With these innovations, intense disagreements emerged and continue. As Bill Bryson quips, "Taxonomy is described sometimes as a science and sometimes as an art, but really it's a battleground."[76]

THE TREE OF LIFE The tree of life emerged as the foundation and emblem for the biological sciences, much like the periodic table in chemistry. While systematists argue about how best to construct phylogenetic trees, there is still a deep impulse for "tree thinking."[77] Phylogeny (a term coined by biologist Ernt Haeckel), or the generation of species, implies that closely related species share more traits with one another. Once species diverge, their evolution takes them along different branches to new adaptations and innovations. As Darwin wrote to a friend, "The time will come I believe, though I shall not live to see it, when we shall have fairly true genealogical trees of each great kingdom of nature."[78] This is always the quest of western science: a singular, universal, and unified truth.

But on what basis does one organize an evolutionary tree? What traits and changes mark evolutionary change? Depending on which traits or combination of traits one selects, the phylogenetic trees that emerge can be wildly different. The field moved from typological thinking, where organisms and groups had "essential natures" and where variation was largely ignored, to a model focused on populations and variation. These ideas—population and variation—are immensely vague and politically malleable categories.

In modern systematics, the most influential figure is the German biologist Willi Hennig, who founded the field of cladistics.[79] Hennig organized groups that shared a common ancestor through the structure of a branching tree.[80] Cladistics is organized into monophyletic groups: the most recent common ancestor and all its descendants. At the base are ancestral organisms that branch (or not) into the tips with extant species. In cladistics terminology, all species have some characteristics that are ancestral (plesiomorphic) or derived (apomorphic) relative to other species. For example, for reptiles, scales and four limbs are ancestral traits. Lizards have both scales and four limbs, whereas snakes have scales (ancestral) but no limbs (derived). This does not imply that one is superior to the other; phylogenetic trees merely capture the acquisition and loss of characteristics, which can happen multiple times. The key point here is that we should not talk about any species that lives today as primitive, ancestral, lower, advanced, or higher. All extant taxa are of the same age, having evolved for the same amount of time from a common ancestor some 3 or 4 billion years old. The tree-thinking view of species is thus quite egalitarian—it organizes life, past and present—and helps us visualize varied evolutionary adaptations and variation but ultimately celebrates all species.

With the advent of cladistics and, more recently, computing power and molecular technologies, evolutionary genetics and the genetic sciences have taken over the field. Genetic data and the discovery of many genes in plants have produced data for new evolutionary trees. But not all scientists are converts. Some taxonomists still work with morphological characters using the power of large databases, a synthesis not possible earlier. Others, unconvinced that genes are the primary site of evolution, continue to ground evolutionary thinking in the adaptations of organisms.

Infrastructures of Naming, Nomenclatures, and Classifications

In their exploration of classification and its importance, Bowker and Starr argue that classifications and standards occupy a critical place in the studies of social order.[81] Modern systematics might reclassify and reshape our understanding of evolution, but much of plant sciences outside academia have practical purposes—agriculture, horticulture, and the worlds of nature lovers and amateur naturalists. Academics and their focus on phylogenetic relationships don't always see eye to eye with applied botanists outside of academia.

In working on this history, I have come to appreciate the vast (yet invisible to most) human infrastructures and standards that keep the botanical world functioning. Embedded in histories of infrastructure are organizational histories of taxonomists and their naming protocols.[82] Classification and standards are closely related but not identical. Over the history of botany, many have proposed classificatory systems, but as we know, most did not survive or become the standard. In some instances, multiple naming systems coexist.

How are nomenclatures regulated? I want to briefly explore the very subjective and deeply Eurocentric roots of nomenclatures that standardize botany. Critically, I want to explore what standardization does, whose standards prevail, whose knowledge is erased, and what is lost. I want to ask how botanical worlds might have been otherwise and how we might access these histories for the future.

Botanical nomenclature has never been fully aligned with botanical classification. Decisions on nomenclatures are not arbitrary but are regulated by the taxonomic community through a "family planning" committee that meets regularly.[83] In 1982, for example, International Botanical Congress in London adopted a set of basic rules. To quote Gledhill, they included the following:

1. One plant species shall have no more than one name.
2. No two plant species shall share the same name.
3. If a plant has two names, the name which is valid shall be that which was the earliest one to be published after 1753.
4. The author's name shall be cited, after the name of the plant, in order

to establish the sense in which the name is used and its priority over other names.[84]

Rules such as these are considered at each International Botanical Congress held every five or six years.[85] The two-century-old effort to establish and stabilize scientific naming of plants has not been easy. In recent years scholars from formerly colonized nations have challenged this system. A lot is at stake. Two examples give a flavor of the debates.

Some systematists in 1999 proposed that the tomato should move from the genus *Lycopersicon* to the genus *Solanum*.[86] While this change may satisfy academics, for seed companies, the costs of relabeling, repackaging, and advertising could run in tens of millions of dollars. A debate ensued. Ultimately, the tomato was renamed from *L. esculentum* to *S. lycopersicum*. If this happens repeatedly, costs can be considerable. For the buyer and seller of tomatoes, does the scientific nomenclature matter? Should it?

More recent is the case of the genus *Acacia*, a massive cosmopolitan genus across Africa, Asia, the Americas, and Australia. Growing genetic evidence suggested that the genus was not monophyletic (that is, descended from a common evolutionary ancestor). Some proposed a split: one lineage in Australia and another in South Africa. But which would retain the name *Acacia*? Which local species would emerge as the type species? Nature and nationalism aligned—both countries claimed the acacias. A passionate and contentious debate ensued with strong political and nationalist fervor. Ultimately, the 2005 International Botanical Congress in Vienna approved a decision to conserve the name *Acacia* for Australia while two pan-tropical lineages were named *Vachellia* and *Senegalia*. Interestingly, this was possible because the committee adopted minority rule even while the majority voted against this decision. Arguing for a procedural violation, this issue was revisited in 2011 but with similar results.[87]

Many professional and amateur botanists in Africa, Asia, and Central America were incensed by the decision, arguing that a well-financed, politicized campaign by Australia was behind it.[88] It ignored the tree's African symbolism. Does retyping the genus signify "yet another act of economics and cultural pillage of the Third World by the First?"[89] The website for the

South African National Parks, under a banner headline "Africa to lose all its Acacias," pointed out that nine people had decided for the entire world. A newspaper in Nairobi expressed outrage in the headline, "Did you know it is illegal to call this tree Acacia? Australia claims exclusive rights to the name."[90] So deep is the opposition that despite the decision, several authors writing about African acacias have pointedly retained the genus name *Acacia*.

This split between academic botany and practitioners in the field repeatedly emerges. For example, gardeners find broad categories of roses, cabbages, carnations, and leeks to be perfectly adequate for their purposes. They are satisfied with a naming system that does not keep up with botanical rules or reassignments based on molecular information. Indeed, many seed and plant catalogs avoid botanical names altogether. Changing names as per cladistics and systematics is annoying and costly. However, if names are not changed when new information is discovered, then naming loses its connection with theory and emerges as an arbitrary mapping of world without any system governing it.[91]

Some worry that an exclusive focus on academic evidence in the name of scientific objectivity denies the feelings for local traditions in plant names, undermining the ability of the *International Code of Botanical Nomenclature* to serve as a universal system.[92] There is value in local knowledge. Sometimes it is more insightful and long lasting. For example, oral traditions in Hawai'i have accurately identified individual algal species, while three-fourths of the scientific names have changed in the past ninety years.[93]

Botany as an Imperial Archive

A central impulse in the history of taxonomy is creating an imperial archive. During colonial times, taxonomy helped consolidate knowledge for a more efficient extractive empire. Today, its goal is to chronicle all life on earth. Biologist E. O. Wilson writes:

> Now it is time to expand laterally to get on with the great Linnaean enterprise and finish mapping the biosphere . . . discover and describe the Earth's species, to complete the framework of classification around which biology is organized, and to use information technology to make this knowledge available around the world.[94]

Bowker characterizes such a view as a "panoptical" dream, an enduring impulse toward an imperial archive.[95] Indeed, many have attempted such inventories.[96] A few examples include the Barcode of Life (BoL), a project literally inspired by supermarket barcodes, and a crowd-sourced inventory by systematists and qualified amateurs to produce a Wikipedia-styled website, Encyclopedia of Life (EoL).[97] Earth BioGenome Project, a multibillion-dollar venture, attempts to catalog the genomes of all species on earth and is labeled by its organizers as a "moonshot for biology." The ensuing sequences will be used to create a Digital Library of Life (Nelson 2018). Likewise, the E. O. Wilson's Biodiversity Foundation's Half-Earth Project aims to create an Inventory of Life.[98] The move to register all names, to agree on model data structures and formats for botanical databases to facilitate biodiversity management, is just as urgent and just as overly optimistic as Alphonse de Candolle's calls in the nineteenth century for a rational system of nomenclature.[99] Such incomplete utopian projects are so pervasive in the history of naming and record-keeping that they should be regarded as standard rather than abnormal.[100]

Despite such initiatives, taxonomists lament the decline of the field, at a time when the "sixth extinction" is upon us.[101] We are losing species much faster than we can discover and name them,[102] and while taxonomists tout abstract notions of chronicling biological, in reality there is inadequate capacity for this. The field remains Eurocentric, with little organizational interface to facilitate a fair exchange of information.[103] There continues to be a "*de facto* information imperialism, causing a net flow of raw data out of the Third World into Western databanks—where it is converted into economically valuable information and knowledge, and then sold back."[104] Like in colonial times, western scientists continue to "parachute" into third world nations, access their botanical resources, and return home to process their data.[105] It is an old problem—third-world countries catching up with first-world technology, and poor community groups bridging the great divide.[106]

The project of decolonization necessitates deep reflection of such ongoing colonial practices. How do we build collaborations between scientists and institutions in formerly colonized worlds? How can international infrastructures undo colonial hierarchies of power? The answers are decidedly complex and rarely satisfying or equitable. For example, although indigenous peoples have used certain plants for centuries, patents still elude them within western legal regimes. At best, it is possible to prevent individual scientists and companies from being granted exclusive patents. Fascinating cases involve indigenous groups, growers, scientists, and multinational companies working to craft workable contracts and profit-sharing agreements. In *Bitter Roots*, Abena Dove Osseo-Asare tells the story of six African plants and the production of pharmaceutical products from them.[107] The indigenous San peoples have long used Hoodia to reduce hunger, increase energy, and ease breastfeeding. In *Reinventing Hoodia*, Laura Foster chronicles how the plant was reinvented through law, science, and the marketplace and collaboratively repackaged as an appetite suppressant.[108] Indigenous peoples, scientists, growers, and multinational companies negotiate contractual terms within uneven political and legal landscapes. While indigenous groups have pushed back to create more equitable arrangements, colonial botany has laid the foundations for its enduring power.

Plant "types," the original specimens used to typify a species, are essential references during identification.[109] As we have seen, types are not always hosted in the country where they were collected and are usually held in western nations. For example, in their analysis of the family *Rubiaceae* in Angola, they found that in 430 taxa (96 percent), the actual type was in European herbaria. Only 50 taxa (12 percent) have type material represented in African herbaria. Colonial legacies endure in their afterlives. Yet there is a deep reluctance to change or undo these legacies.[110]

In the impulse to decolonize, should materials be repatriated? Decision makers and politicians prefer repatriating digital images and databases rather than the physical archive.[111] Ironically, at the same time we have seen a disinvestment in western archives and a decline in the number of taxonomists working in those archives. Young African taxonomists are frustrated about not being able to access, let alone maintain and expand, their work while European institutions grow moribund without functioning budgets or staff, "where valuable African specimens are deposited, uncurated and unused, running the risk of slow, inevitable decline, if not decomposition."[112] In short, postcolonial nations that have invested in expertise have no access to control the representations of flora of their own nations. An exception is an initiative, Flora of Ethiopia, where external research funding has worked alongside building of local infrastructures.[113]

The Rhodes Must Fall campaign, which started in Cape Town, South Africa, in 2015 and then spread across the world, has added urgency to these questions. Cecil John Rhodes, a colonist, imperialist, and racist and "a symbol of the ruthless imperialism," has 126 plants (collected between 1899 and 1976) that bear his name. The name of Rhodes has been erased from the names of nations (Rhodesia), and the recent South African campaign challenged the presence of his legacy in the statues and in the names of institutions. Yet some activists astutely ask: why is he still celebrated in the botanical world?[114] Rhodes is, of course, only one of many colonists who continue to be honored in the botanical imagination. During colonial rule, white colonists replaced local names with those of royal figures and the names of white botanists. Why should those names remain? Why should botany reinforce colonialism through a parade of western names? With the rise of digital media, change should be easy.

Many environmental organizations such as Sierra Club and the Audubon Society are reckoning with the nativist and racist beliefs of their founders. The continued link of environmental groups to nativist and xenophobic ideas, dubbed the "greening of hate," is a reminder of enduring racist and colonial legacies.[115] While the national chapter of the Audubon Society has retained its founder's name, some local chapters have opted for new names.[116] The American Ornithological Society announced on November 1, 2023, that it would change the names of all bird species named after people. Further, it committed to involve a diverse representation of experts and the public in the process.[117]

Questions of control of botanical infrastructures also pose a problem. For example, botanical repositories in the planet's tropical, subtropical, and temperate regions house a wealth of global flora. Yet, as we have seen in the case of *Acacia*, those countries hold only a fraction of the votes necessary to amend the *International Code of Botanical Nomenclature*. Colonial-era rules have proved difficult to change given the enduring coloniality of power.[118] Today, plant names, nomenclatures, and taxonomies are all abstract, scientific exercises. No doubt they are useful and important to the life sciences, but they contain less utility for the nonspecialist plant lover. One of the problems with the taxonomic system is that it is not designed to be an information retrieval device. In folk taxonomies, names are practical and pragmatic—they provide contextual information to others who already share an understanding of the organism's cultural significance.[119]

It is important to note that other imaginations in botanical history offer alternatives. For example, Sir William Jones, an oriental scholar and linguist, developed a great interest in botany. As a philologist, he used indigenous knowledges and languages.[120] He drew often on Sanskrit names for plants, arguing, "I am very solicitous to give Indian plants their true *Indian* appellations."[121] He gave plants a Sanskrit name, a uninomial, followed beneath by its common name, and beneath these its Linnaean or other scientific equivalent. In part the local Indian languages were provided to make resource identification and extraction more efficient.[122] Alas, he published only two papers on Indian plants, both posthumously.[123]

Plant sciences focus on organisms that bring funding. Not all organisms receive equal attention, even commensurate with their importance. For example, there are more viruses than bacteria in the open ocean.[124] But viruses, though plentiful, do not get as much attention as bacteria. Similarly, there is a bias toward flowering plants, especially those that are big, exotic looking, tasty, or economically important. Many of the most urbanized coastal areas where human impact has been greatest have never been surveyed.[125] An analysis of plant protection legislation in twenty-nine European countries lists taxa almost entirely composed of vascular plants and dominated by spectacular species attractive to collectors and the public.[126] Despite claims of "objective knowledge," western botany retains its fascination with the exotic other.[127]

THE COLONIALITY OF POWER

Sylvia Wynter, drawing on Aníbal Quijano, Walter Mignolo, and Michel Foucault among others, offers a foundational story of colonialism.[128] Colonial knowledge systems, she argues, transformed the human into Man. The vast intellectual enterprise of western civilization of "Man" should not be confused with the human or humanity. To unsettle the coloniality of power, she argues, it is critical to unsettle the overrepresentation of Man, a central secular figure of modernity. Wynter identifies two historical stages—the first from the Renaissance to the eighteenth century, when Man was secularized into a rational political subject (Man1, *Homo religiousus* or *Homo politicus*), and the second from the eighteenth century until today, which captures the Darwinian vision of the naturally selected Europeans (Man2, *Homo oeconomicus*). *Homo reli-*

giousus emerged from a "theological order of knowledge of pre-Renaissance Latin-Christian Medieval Europe," a figure that ushered in *homo oeconomicus*, a bourgeois figure of liberal mono-humanism from the late nineteenth century. Man1 makes possible the development of the physical sciences, and Man2 the rise of the biological sciences.[129]

Man2 is the figure of the colonial botanist. To understand why this is important, Wynter narrates theology—the Christian God was reconceived as a caring father who created the world/universe for mankind's sake. This made way for Copernicus's declaration that since the universe was made for our sake, it was knowable. This belief undergirds the power of botany, plant naming. The shift from divine providence and/or retribution, and from witchcraft and sorcery, to the new principles and laws of nature ushered in modern science and medicine. For the history of botany, it is important to recognize that "nature" became the "representative agent" of God on earth. This "nature" is the object of scientific inquiry.

I spend time discussing coloniality for several reasons. First, it helps us understand how colonialism is at the root of modern science and its rise. Second, it helps explain why and how the western human created hierarchies of being—racial and gender hierarchies in humans and classification hierarchies among the nonhuman. Plants, seen as nonsentient and immobile, were at the bottom of such hierarchies and thus were turned into objects of science and commerce. Among humans, European men were considered superior, followed by white women and other humans (if indeed they were considered human at all). Third, it's important to understand coloniality because the sciences birthed from colonialism inherited many of its characters—an abstract "natural" world for human consumption, the idea of a disembodied scientist, the practices of an objective scientific culture, and finally the supremacy of science and the scientist as the site of knowledge. Finally, colonialism and colonial ideology knit the world together through academic knowledge. While we inhabit disciplinary silos today, they are connected historically. The history of botany is not a singular history of plants but a colonial worldview and ideology imposed upon all who were not the figure of "Man." Any project of decolonizing botany must simultaneously also decolonize a worldview and ideology. The history of botany is a story about more than plant worlds.

I began this chapter discussing how naming and organizing living beings were meant to ensure *order* in the world. Theories of evolution and DNA technologies have transformed how we understand life on earth. The idea of the tree remains central to the evolutionary imagination, a motif for the branching and deeply rooted connectedness of all life on earth. At the end of the branches are discrete species of life. In *Imperial Leather*, Anne McClintock powerfully unpacks two central genealogies of imperial nature: the combined figures of the tree of life and the white "family of man."[130] Central to this book is how gender and race traverse species lines to ground hierarchies of difference of both the social and biological, the human and nonhuman alike. In short, the story of "man," where hierarchies of race and gender are saturated with meaning, is also the story of the nonhuman. Colonial and repressive actions often appeal to both ideas of nature and the family.

By now rich accounts of colonial botany have persuasively argued that imperialism has profoundly shaped our view of nature—how we know nature, our relationship to nature, the individuals who produce knowledge, and our theories of the natural world.[131]

The figure of the tree has transformed as new technologies have arisen. Human phylogenetic trees visualize kinship not between species but between human "populations." The tree of life's representation of evolution has shifted from an earlier hierarchical model of evolution as a ladder to a branching structure of partition and diversification, although trees as a motif tend to downplay genetic variation within species. One limiting factor of the tree is that the branches can never merge. Here, histories of biology inspire. Julian Huxley, challenging racialized views within biology in 1936, argued that unlike plant and animal species, there was too much interbreeding throughout human history for speciation to occur or for there to exist isolated and unique populations.[132] Instead of a branching tree, Huxley introduced the idea of reticulate evolution where branches merge and join each other—a network rather than a tree.

The diverging tree is embedded in evolutionary and popular imagination with a singular focus on biological difference, most prominently gender and racial differences.[133] The Human Genome Diversity Project as started in the

1990s with the idea that genetics could help us understand human migration and evolution. Jenny Reardon powerfully shows how the World Council of Indigenous Peoples saw this approach as fundamentally racist and colonial—a "vampire project."[134] As Reardon argues, population genetic studies, like all scientific studies, are never "pure" but always a coproduction of the social and natural order. Similarly, Lisa Gannett reminds us that ideas about the biological concept of "race," while ostensibly rejected after World War II, never disappeared but were recoded in the terminology of "population" thinking. Powerful colonial ideas and tropes—like gender and race—have afterlives because they became naturalized as scientific concepts. Race gets increasingly codified within evolutionary language rather than ecological ones, perpetuating the language of fundamental genetic differences among races.[135] In this view, we are not our environment, we are our genes. These tensions and representations of evolution also shape our view of plants. Given this racialized history, should we rethink the very idea of species?[136]

Huxley's proposal of reticulated evolution has reemerged in recent years as a model for life on earth. Interbreeding between one branch and another can add layers of complexity. In addition, works by Lynn Margulis, Tsutomu Watanabe, Carl Woese, and W. F. Doolittle have challenged the exclusive focus on vertical inheritance across generations and instead highlighted the lateral movement of genetic material across species lines—that is, horizontal gene transfer.[137] Barbara McClintock's work has also highlighted the movement of transposable elements, a process that challenges simplistic Mendelian inheritance. Roving genes have now been found in every branch of the tree of life where geneticists have looked: "It's not just rare freaks or accidents, it's happening all the time. And in quite divergent species too."[138] This movement of DNA across species and across time fundamentally challenges any neat phylogenetic tree and also the idea of biological individuality.[139] If organisms transfer and accept DNA across species within a generation, how unique is the individual? In the field of human paleoanthropology, some scientists have suggested a metaphor of a braided stream, where streams can branch out but also flow back into the main river.[140] Recent work suggests that we should move away from the tree to a network or reticulated evolution.[141]

I began with the messy world of postcolonial naming—naming and renaming, multiple namings, sedimented meanings. Models of reticulated evolution in a profoundly naturecultural world inspire the title of this chapter as the "categorical impurative." Throughout botany's history, scientists have attempted to come up with a definitive genealogical tree of life, a goal that proves elusive. Emily Singer argues that "the current version of the tree of life is more like a contentious wiki page than a published book, with certain branches subject to frequent debate."[142] Indeed, one geneticist, after a study of yeast populations, concluded, "We are trying to figure out the phylogenetic relationships of 1.8 million species and can't even sort out 20 types of yeast!"[143] As the history of nomenclature and classification has taught us, the problem is not the lack of the right technique or biological variable or theory. The problem is the imperial quest for a universal, unitary, pure genealogy of life on earth. Perhaps there is no unitary order? Perhaps we should put quests of purity aside to embrace the ebullience of plant adaptations, the diversity of wondrous morphologies, and the messiness of life on earth. Embrace the categorical impurative!

CHAPTER FOUR

Perhaps the World Ends Here

Spicy Embranglements in the Postcolony

But in truth, should I meet with gold or spices in great quantity,
I shall remain till I collect as many as possible, for this
purpose I am proceeding solely in quest of them.
CHRISTOPHER COLUMBUS | *Journal*, 1492

Each spice has a special day to it. For turmeric
it is Sunday, when light drips fat and butter-colored
into the bins to be soaked up glowing, when you
pray to the nine planets for love and luck.
CHITRA BANERJEE DIVAKARUNI | *The Mistress of Spices*

NO DISCUSSION OF SOUTH ASIA IS POSSIBLE WITHOUT FOOD. No discussion of food is possible without plants. No discussion of plants is possible without colonialism. The colonization of South Asia—its food, plants, and knowledge systems, and the ensuing global trade, especially of spices—profoundly shaped European colonialism.[1] As Columbus describes it, the value of spices was equivalent to gold. The very euphemisms that describe this brutal historical period as the "Age of Exploration" and the "Age of Discovery" speak volumes. But botanical trade, especially the spice trade, has a much longer history. Global demand for spices grew throughout the Roman era and the medieval period, redrawing the world map and defining economies from India to Europe.[2] Those who controlled spices controlled the flow of wealth in the world. India, historically part of South Asia, is a land that brought many to its shores—Persians in 500 BCE, Greeks in 150 BCE, Arab traders in 712 CE. Colonists particularly sought black pepper, cinnamon, turmeric, clove, and coriander. Rich resources brought traders, then empire, and then colonial rule where botanical commerce propelled the growth of capitalism.[3]

While Romans and Arab traders had a long tradition of trading with South Asia, the story told of colonialism usually begins with the Portuguese, specifically Vasco da Gama's arrival to the Malabar Coast in 1498, creating a spice route from South Asia to Europe. Subsequently, the British came in 1610 and ruled the country for over three hundred years. At first glance, fertile lands that yield bountiful spices alongside a richness of culture should be a prized asset, something to be celebrated. Yet history teaches us that such bountiful riches on a land were in fact a "resource curse," attracting colonists to the shores, followed by brutal regimes of colonial rule to exploit and control the resource.[4] It was lucrative business. For example, as Amitav Ghosh chronicles, in the "Banda Islands, ten pounds of nutmeg cost less than one English penny. In London, that same spice sold for more than £2.10s.—a mark-up of a staggering 60,000 per cent. A small sackful was enough to set a man up for life, buying him a gabled dwelling in Holborn and a servant to attend to his needs."[5] This legacy of food and spices shaped and was shaped by the science of botany, and its legacies endure in contemporary plant sciences.

This chapter traces colonialism and its botanical legacies with a focus on South Asia. Colonial exploration and exploitation necessitated control of nomenclature, the very system that continues to ground modern globalized trade.[6] To illustrate the centrality of norms and nomenclature, I juxtapose two historical moments that demonstrate the power of food, colonialism, and capitalism: first, the making of the treatise *Hortus Malabaricus* in the late seventeenth century, which compiled a list of plants of the Malabar region; and second, the Traditional Knowledge Database Library (TKDL) in the twenty-first century, which compiles details of traditional knowledge of the region. The importance of plants and their medicinal and therapeutic values, then and now, reveals the heavy and enduring hand of colonialism. These cases make visible the high stakes of liberal logics in botanical language and nomenclature, the *coloniality of nomenclature*, in shaping colonial knowledge practices and their postcolonial and neocolonial legacies.

At the Kitchen Table

Let us start in the kitchen. I grew up in urban India in densely populated flats. When you stepped out of the house each morning, the wafting aroma of food filled the senses. Walking by, you could often guess the household's menu for the day. Every dish has a different combination of spices. You can smell it, taste it. The rhythm of the South Asian kitchen is centered around spices. Glancing at the spice bottles in my pantry, I see a cornucopia of tastes and smells. Most Indians know the names of spices in multiple languages, comparing and trading names as we savor the complexities of regional cuisines. Spices, for their best flavor, need to be fresh, appropriately prepped, and then added in studied order and proportions. But food in the South Asian household is never only about sustenance; every spice has its curative properties. Food and medicine have a tightly knit history. Many of what we consider quintessential Indian spices came to South Asia's shores during the global trade in spices. Indeed, during a traditional *shraddham*, a ceremony performed to honor a dead ancestor, only vegetables purportedly used in ancient times before global trade are permitted. For example, heat can be from black pepper but not green chilies and sweetness from jaggery but not sugar; some vegetables and spices are permitted but not others. I begin with this description to record the deep and varied vegetal registers of South Asia, old and new.

Joy Harjo's poem "Perhaps the World Ends Here," which provides the title for this chapter, opens with, "The world begins at a kitchen table. No matter what, we must eat to live." The poem then goes on to describe the naturecultural life at a kitchen table—the vegetables, animals, beverages, teething babies, conversations, gossip, laughter, lovers, life, joy and sorrow, sun and rain, life and death, and gratitude. She ends, "Perhaps the world will end at the kitchen table, while we are laughing and crying, eating of the last sweet bite." Harjo captures the centrality of food, its naturecultural dimensions, the cooking and eating, as the heart of human sociality.

The kitchen remains a gendered space. While famous commercial kitchens remain masculine spaces, the kitchen of the home endures as feminine space, both in its sexist registers of domesticity and also as a reclaimed feminist space of resistance and power. Indeed, it is no accident that the influential women of color publisher in the 1980s was named Kitchen Table: Women of Color

Press. Through the kitchen table, we can tell the complex stories of plants and their spicy embranglements with the histories of sex, gender, race, caste, class, sexuality, and nation.

Names, Norms, Nomenclatures: Tracking Colonialism and Its Legacies

As we saw in the last chapter, plant nomenclature has a complex history. While local plant names may persist in the colloquial, in the world of botany, one needs to learn their Latin names and nomenclature. It was well into my forays into the history of science that I discovered that these names were sometimes those of colonial botanists and sometimes latinized local names. The binomial nomenclature system offered a universal language of botany, critical for the extractive logics of empire.[7] It allowed a standardization of language across colonies, the marking and branding of plant conquest, and more importantly the creation of imperial botanical indices and maps. Plant explorers sent out to the colonies returned with precise knowledge and often with specimen plants.

While an embrace of a universal botanical nomenclature might have aided imperial ambitions, it has marked the plant world in enduring legacies of "plant colonialism." I focus on plant language and nomenclature to explore two cases: *Hortus Indicus Malabaricus* (henceforth *Hortus*) in the seventeenth century and the Traditional Knowledge Digital Library (TKDL) in the twenty-first century, which serve as bookends of the legacies of colonial botany.[8]

THE CASE OF THE *HORTUS MALABARICUS*

Hortus Malabaricus ("Garden of Malabar"), a comprehensive botanical treatise that chronicles the flora of the Indian region of Malabar, was the brainchild of the Dutch governor of the Malabar, Hendrik van Rheede. The oldest comprehensive published work on tropical and Asian plants, the *Hortus* was written in Latin and published between 1678 to 1693. Luxurious life histories and the community of plants are reduced and represented in the lexica of colonial botany. The *Hortus* is the culmination of thirty years of collecting, compiling, and editing.[9] Interestingly, despite much talk about Indian biodiversity and indigenous medical system, the *Hortus* has not been updated, and over three centuries later it remains unsurpassed in the importance and magnitude of its presentation of the medicinal plants of the Malabar region.[10] Equally remarkable is that there was no translation available until the annotated English edition in 2003, translated by K. S. Manilal, a botany scholar and taxonomist who devoted thirty-five years of his life to researching, translating, and annotating the *Hortus*.[11]

To understand the development of the *Hortus*, one needs to understand Dutch colonialism. The rich resources of spices and plant lives brought the colonizers to the Malabar region in the southwest of India, well known for its spices and medicinal plants. The Malabar coast has long been important to foreign commodity traders, from Arab traders of thirteen hundred years ago to the present day.[12] According to Hendrik van Rheede, the main reason for the *Hortus* was to help the Dutch East India company. Living in Cochin, Rheede observed and recognized that locals had a trove of plant-based medicines to cure illnesses, sources that Europeans in the region also began to use. However, when these medicines were sent to Europe by sea, they often arrived decayed and spoiled, with little value or use. Rheede was prescient and recognized that a catalog of Malabar's plants that included their curative properties "would involve great profit."[13] Indeed, the majority of plants included in the *Hortus* were regarded as medicinal plants, serving as a source of medical prescriptions for over two hundred diseases. While these were diseases prevalent from the fifteenth to nineteenth centuries, the plants remain important to this day.[14]

Rheede's vision paid off. As Europe saw a steady influx of exotic plants into the continent, botanists in the seventeenth century grew interested in how to classify and catalog them. In the meantime, European botanical gardens, notably the Amsterdam Medical Garden, became havens for exotic plants. Rheede's *Hortus* fed the European appetite for colonial exotica and translated orientalist and exotic visions of the east into European lexica. Overall, the *Hortus* was a huge success and regarded as the best and most comprehensive source for information about the flora in India, especially the Malabar region. It was used extensively by leading botanists, including Linnaeus. In fact, Linnaeus celebrated Rheede by naming a genus after him, *Rheedia*.[15] Among the numerous books that Linnaeus studied thoroughly before producing *Species Plantarum*, the *Hortus* was one of two that he singled out for respect. Linnaeus included 258 Malayalam names of plants from *Hortus Malabaricus* in *Species Plantarum*, adopting many Malayalam plant names to coin binomials directly or after Latinizing them. Indeed, K. S. Manilal states that of all the plant names derived from Indian languages in *Species Plantarum*, the largest number are of Malayalam origin.[16]

How did Rheede conceptualize the *Hortus* and its knowledge base? He clearly relied on the expertise of local people of Malabar, which was based on their pre-Ayurvedic traditional knowledge.[17] But the *Hortus* is a pre-Linnaean work, and there are no known specimens or even exact scientific identities of the plants illustrated. Within contemporary botanical knowledge, how accurate is the *Hortus*? This question has been much debated. Since the plants represented in the *Hortus* cannot be carefully checked against preserved specimens on herbarium sheets, it has raised much scientific skepticism. Yet Linnaeus and others after him used the illustrations and descriptions from *Hortus* as "types" for their classifications. In fact, they do not often correlate with contemporary botany.[18] Historians who have studied the development of the *Hortus* argue that it is clear that local Indian scholars selected the plants and described, collected, and compiled their knowledge of their medical properties.[19] As Fournier argues, Rheede was no botanist, but he was a "successful organizer."[20] Also evident is the fact that during the time of the plant collections, the King of Cochin was not on friendly terms with Rheede, and as a result Rheede could not move about freely to supervise or corroborate the collections. Scholars conjecture that only volumes 3-8 were personally "scrutinized" by him.[21] Supporting this picture is the fact that Rheede expressed fear that the *Hortus* would be marginalized because it relied so heavily on local knowledges and terms that were unfamiliar to "learned botanists" and scientific usage of that time. To add to this complexity, what we see in the *Hortus* was no easy work of translation. To illustrate this point, consider this description of how the *Hortus* came into being:

> The vast majority of Malayalam words and expressions had (and still have) two forms, more or less different from each other, namely the spoken form (*vaamozhi*) and the written form (*varamozhi*). The local names of plants were dictated to Rheede in the spoken (*vaamozhi*) form of the local language Malayalam. That was then translated to Portuguese language,

> writing the names in the spoken form of the names themselves, from which it was translated to Dutch language and then from Dutch to the Latin language used in the printed text of the book. During this [*sic*] tortuous, multi-stage transformations, the Malayalam names which even otherwise do not easily yield to European tongue or ears, had undergone severe distortions, that are reflected in their depiction in Roman script in the book. Alongside the illustrations, Malayalam names are written in Malayalam script also, which is a great help in understanding them better. However, they are the exact transliteration of the spoken form (*vaamozhi*) that was in use in the 17th century, written in an old script. Since then, Malayalam script itself as well as the language have changed. To compound the confusion, it also appears that many of the names in Malayalam script are written, perhaps later in Amsterdam and Leiden, by persons who had no knowledge of the language or its script.[22]

A convoluted process! So much is lost in translation. We see the erasure of the many laboring bodies that helped produce this treatise. We see class and colonial privilege where only the colonists and upper-caste Indians are named (with one exception). Despite this complex history, what we see in the main colonial account is one that is entirely hagiographic, with Rheede as the primary architect and hero of the *Hortus*. Rheede is the one who is credited with organizing a varied assortment of people—Indians and Europeans, botanists, priests, and clerks, medics and physicians, and soldiers—working across two different continents (at that time when the continents were half a year's journey apart) to create the twelve-volume set over a period of thirty years. Similarly, Kapil Raj discusses the French herbal chronicles in Orissa at the end of the seventeenth century, which drew on fakirs, male and female collectors, illustrators, translators, bookbinders, and mediators.[23] In the colonial archive, the role of local Indians is lost, and the few recorded remain at best minor figures. No women are named in this history. What is celebrated is the power of colonial extraction and appropriation and the scientific prowess of the colonial forces.

Only in recent times has the narrative of Rheede as the sole hero of the *Hortus* been contested, starting notably with the environmental historian Richard Grove, who argues that the *Hortus* was based on the Ezhava (considered a lower caste from Kerala) system of botanical knowledge and classification.[24] The Ezhavas of contemporary Kerala as a group are regarded as belonging to an OBC (Other Backward Class), a terrible but official term used by the government of India to classify castes that have been discriminated against, and educationally and socially disadvantaged.[25] Of the large numbers of Indians who worked on the *Hortus*, the key figures credited include three Brahmins, Ranga Bhat, Vinayak Pandit, and Appu Bhat, and a Malayali and Ezhava physician, Itty Achuthan.[26]

Historians agree that Rheede relied on local knowledge in constructing the *Hortus*. Itty Achuthan, a Chogan (Ezhava), belonged to a family who had been physicians for generations. As Burton Cleetus argues in an insightful essay, Ezhava physicians in precolonial times had widespread and extensive knowledge of the medicinal value of local flora and fauna.[27] This knowledge included herbal remedies as well as other healing practices such as rituals, incantations, and spells. While Itty Achuthan was not given authorship, his testimony is included in the preface of the *Hortus*, where he clearly claims his botanical knowledge and medical expertise developed through "long experience and practice."[28] However, historians cite many other cases where local experts are not named at all in the official record, and therefore all evidence of their contribution is completely erased from the record.[29]

With the arrival and consolidation of British colonial rule, new issues arose. Contestations and consolidations emerged between the British and Indians, as did contestations within Hindu groups, especially along caste lines. By the mid-nineteenth century, Ezhavas and other lower-caste communities began to challenge the widespread caste discrimination. This led to large-scale communal clashes and social tensions. By the mid-nineteenth century, the ruler of Travancore ordered the opening up of formal Ayurvedic education to lower castes. We begin to see the dissemination of classical knowledge and other art forms among the Ezhavas, who then spread this training among fellow caste men. As a result, we see how upper-caste Ayurvedic texts negotiated colonial medicine and how these negotiations and hybrid knowledges were spread to lower-caste and local indigenous medical systems. Cleetus argues that this process can be characterized as a "hegemonisation attempt by the dominant tradition over the lower castes."[30] Recent work on the *Hortus* is also a reminder about the contested nature of history. As postcolonial historians have revisited the history of the *Hortus*, we see a distinct blurring of boundaries between western and indigenous medicine and between upper- and lower-caste knowledge in practice, even though these categories have been consolidated as pure histories. We have evidence of exchange between European and Ayurvedic medical systems through Jesuit missionaries from as early as the early sixteenth century.[31] However, indigenous societies remained deeply heterogeneous in their culture, geography, sociality, and health care methods and practices.

As the Indian struggle for independence grew, Hindu intellectuals drew pride from the ancient Vedas *as* Hindu and science. Modern sciences, they claimed, were an extension of ancient Vedic sciences. With this began an enduring history where Hindu superiority was grounded in a grand Vedic civilization—a compelling trope of the nationalist imagination.[32] Within the realm of health and medicine in particular, Hindu science was consolidated in the name of Ayurveda (among other indigenous forms), which was often represented as classical and upper-caste male knowledge systems grounded in ancient Vedic knowledge. The nationalist struggle and imagination largely remained in the purview of upper-caste men. Leena Abraham shows how male members of households gained entry into medical practices, while women were relegated into specializations like midwifery, which was regarded as lower in status and often reserved specifically for women from "untouchable" castes.[33] A key rationale for excluding women from the practice was their purported "impurity" during menstruation—menstruation being regarded as a mechanism of cleansing the body.[34] The complex politics of purity—of science, of caste, of nation—endures in the bodies of women. The entry of women into Ayurvedic colleges in large numbers only became possible with the rise of biomedicine and the diminished social status of Ayurveda.[35]

The *Hortus* illustrates some key lessons from the postcolonial and subaltern histories of science. Rather than a model of passive diffusion or wholesale imposition of western science into the colonies, science in India progressed through "negotiations" and a process of "hybridization." By now, historians of colonial science have amply demonstrated that we can no longer assert a "western" genealogy for modern technosciences. Instead, technosciences were already global by the nineteenth century.[36] Colonial science in India tugged it away from its western roots and combined it with Indian, specifically Hindu, forms of knowledge.[37] During the nineteenth and twentieth centuries, Indian scientists and intellectuals constructed their own brand of modernity by selectively incorporating Hindu practices and traditions along with western science. This incorporation of both elements generated its own dilemmas, and these conflicting and contradictory attitudes remain unreconciled and thoroughly alive in hybrid knowledge practices.[38] For example, during the recent pandemic, government officials promoted both vaccines and the therapeutic use of cow urine.[39] India's modernity, with science at its core, was thus at once both Indian and Western. Through the *Hortus*, we see an example of how hybrid knowledges are formed. All the plant names used in the *Hortus* are based on plant names used by local peoples and do not correspond to the more "medical" Sanskrit names used in Ayurveda. As Manilal and Remesh argue, not all the plants mentioned in the *Hortus* are used in Ayurveda, which suggests that the sources for the names were local and not Ayurvedic texts (although as we see above, what is authentically Ayurvedic remains decidedly murky).[40] It is believed that the ethno-medical information of the *Hortus* was culled from palm leaf manuscripts by Itty Achuthan, adding strong evidence that the ethnobotanical and ethnomedical uses captured in the *Hortus* reflect generations of empirical knowledge of the Ezhava community.[41] Manilal writes that the palm leaves and other original sources of the *Hortus* were burned in a fire and have not survived. Thus, ironically, the *Hortus* remains the only "authentic" written (that is, excluding oral traditions) record of this indigenous knowledge system available.[42]

Kapil Raj invokes a relational narrative involving circulations, encounters, interactions, and connections that helps put non-European actors back into the historical story as active participants in the knowledge-making process and restores their agency. The language of circulation decenters the "European great-man, heroic image on which the Scientific Revolution narrative is constructed as a singular European achievement, to the exclusion of all other peoples and cultures."[43] However, such narratives also depoliticize the violence of colonialism, as though these are equal actors trading. While the *Hortus*, in this retelling, displaces Raj's "European great man," the "man" remains solidly in place.

The politics of the *Hortus* endures. As social movements against caste discrimination have gained momentum, indigenous Ezhava groups have claimed the *Hortus* as *their* indigenous property, challenging the Indian state and its claims of the *Hortus* as Indian.[44] They demand that all sources of the *Hortus* be repatriated.[45] While the work remains an important historical record of plants, botanists such as Manilal underscore the importance of its taxonomy and utility. Rheede developed the *Hortus* to help the colonists extract resources of the Malabar region more efficiently, yet it is used as a resource today to document Indian herbs and plant-based medicines because it demonstrates the use of plant knowledge in seventeenth-century India, thus supporting contemporary claims of biopiracy and the harms of unequal globalization.[46] Such a circulation model of knowledge production also imposes liberal values of commerce and exchange.[47]

THE CASE OF THE TRADITIONAL KNOWLEDGE DATABASE LIBRARY (TKDL)

The Traditional Knowledge Database Library (TKDL) is an Indian database that collects the use of "traditional" and "indigenous" plant-based remedies, historically or in contemporary times. The TKDL emerged as India's attempt to address the colonial histories of plant-based pharmaceuticals and modern biopiracy.[48] Biopiracy is when indigenous knowledge of nature, originating with indigenous people, is appropriated and used for profit, without proper permission from, compensation to, or recognition of the indigenous people themselves.[49]

The impetus for the database emerged in the 1990s when India experienced a number of patent challenge attempts by scientists in the west. Particularly concerning was the fact that many of the patents were for plants that had a long tradition of medicinal use in India. For example, in 1995 two emigrant Indians in the United States patented using turmeric as a healing agent for wounds. India issued a legal challenge by documenting turmeric's extensive prior use in India, and the patent was revoked in 1997. The most influential was the case of an attempted patenting of the neem plant for its antifungal properties in 1994.[50] The patent was seen as particularly egregious because the neem is considered sacred in much of India and is associated with a number of divinities.[51] International NGOs and representatives of Indian farmers filed legal cases, and the patent was reversed in 2005.[52] Yet despite the fact that both patents were ultimately reversed, for a poor country like India, it is worth noting that fighting the legal case took over eight years and cost hundreds of thousands of dollars.[53] With time, the Indian government became involved with expensive litigation in US courts over patents given to traditional medicinal knowledge from India, such as turmeric, neem, Darjeeling tea, basmati rice, and yoga positions. In the year 2007, 130 patents and 1000 trademarks were issued for yoga postures and products in the United States![54] In each of these cases, we see how local and contextual knowledges, despite a long history, were translated into reductionist notions of active ingredients.

As such cases proliferated, concern and fears of biopiracy grew. Multinational companies continue to mine the world, especially communities with rich indigenous traditions, for new drugs as the global demand for "herbal" and "alternative" medicines grows.[55] Worldwide, 70 percent of herbal drugs are believed to have been drawn from indigenous medicine, as are most synthetic analogues derived from compounds isolated from plants.[56] The lucrative world market for herbal knowledge has only been growing.[57] TKDL was developed in response to this. A researcher stated,

> Our aim is to make traditional Indian medical knowledge accessible in order to protect it. Our digital library is a collective resource in the management of intellectual property rights. India has had very negative experiences with the patenting of biological resources or preparations copying the recipes of our traditional medicines. What we do is make this classical Indian medical knowledge accessible to hinder its misappropriation. But if you want patent examiners to take this prior knowledge into account when looking at applications claiming the novelty of therapeutic formulations, you have to make it comprehensible, you have to translate it into a language they understand. This is what the Indian government has mandated us to do.[58]

Two global agreements frame such patents. The first, the Convention on Biological Diversity (CBD, 1993), gives nations sovereignty over their own biological resources. The second, the Trade Related Aspects of Intellectual Property Rights (TRIPS, 1995), does not recognize such sovereignty but gives rights based on intellectual property and protection. According to the latter, if communities can prove prior use and knowledge, patents cannot be granted. But this knowledge needs to be recorded and vetted. So even while the knowledge might have been practiced for centuries, there needs to be documentation, and the practices need to be substantiated by these new global standards. The TKDL emerged as a way to set standards and document traditional knowledge in India. The granting of patents for turmeric, basmati rice, and neem in the United States created an urgent situation, and the government of India formed a task force that led to the TKDL. By mining ancient texts and literatures in India's multiple languages—Sanskrit, Hindi, Arabic, Persian, Urdu, and Tamil—the database has chronicled the use of plants in Indian history. It is available in English, German, Japanese, Spanish, and French, and it documents evidence of "prior art" when patent applications are filed.[59] Thus, if patents are filed in other countries, the database serves as a resource in the patent-granting process. As the TKDL summarizes:

> Concrete steps have been taken to prepare a programme aimed at documenting the knowledge and information available in the ancient texts of Ayurveda, Siddha and Unani, etc. and prepare a database on the medicinal plants involved and their medical use. This is expected to prevent remedies based upon the medicinal uses of plants being treated as an invention or a discovery. The TKDL is being developed in the first instance for Ayurvedic and formulations documented in important ancient texts along with the medicinal use therein are being developed.[60]

But of course, as the TKDL has unfolded, complexities have multiplied. Much is lost in these representational and translational processes. The TKDL is designed to be searchable through botanical categories and not through the process of preparation or chemicals.[61] In setting up the TKDL, myriad confounding issues have emerged. Thomas chronicles some of the key issues.[62] First, the politics of nomenclature: What gets to be called indigenous or traditional? While global politics enables such a database, it is tough to date "indigeneity." In a country like India, with numerous migrations and colonial regimes and with many unrelated and unconnected countries and cultures, there is little that remains "pure" or untouched by the histories of the colonial. India itself is a postcolonial invention after the exit of the British. In order to establish a TKDL, the idea of the indigenous needs to be defined and fixed.

We also see how the mantle of traditional knowledge often legitimizes dominant traditions and social hierarchies, further marginalizing already marginal local communities and gender groups. For example, should the *Hortus* belong to India, Kerala, or the Ezhavas? Not surprisingly, dominant traditions by majority and powerful groups are given more attention than the traditions of politically more marginal groups.

Given the histories of colonialism, globalization and global circulations are hardly new. When traditions travel, how should we regard them? As Thomas asks, "Is there a difference in doing yoga in a salubrious suburb in Santa Barbara as opposed to doing it in a little village in rural Gujarat where the practice of yoga is inspired by religion and is part and parcel of a way of life?"[63] Let's also remember that Ayurveda and yoga have predominantly male public faces, although women practice them in equal numbers. Abraham, for example, notes the high rate of women's entry into the traditionally male-centered fields of Ayurveda in recent years but also notes an absence of women, to varying degrees, in the "globalization of Ayurveda."[64] Such examples show the uneven transformations in the inclusion of women into Ayurvedic practice.

The patenting of plants and practices by the west is no doubt deeply problematic, but in creating TKDL as a site of authenticity, new quandaries arise. As postcolonial studies reminds us, science in India is best characterized as a site of hybridity,[65] and yet the TKDL doubles down on global legal frameworks that enshrine authenticity and essentialist claims. Within global regimes of patents, complex histories are reduced to simple representational claims and easy translations. After all, systems like Ayurveda have unique epistemological claims (incommensurable with western medicine) and locate health within larger ecological, social, and spiritual contexts. They have holistic and individualized approaches to diseases. Unlike western medicine, they do not reduce it to a single problem or chemical solution.[66]

The digitization of traditional knowledge is also significant "precisely because there are limits to the digitization of complexity," where TKDL is more accurately an "abstraction of traditional practices from the larger meaning systems that suture and ground a given knowledge and practice."[67] The TKDL thus tries to both protect traditional medicine and interface with global property rights and capitalist biosciences. But what does indigeneity or tradition mean in a history of global circulations and translations? The lessons of postcolonial science and technology studies teach us that despite representations of science as authentic and western, it has been and is translated and hybridized into braided sciences.[68] The TKDL continues this mistranslation of complexity by embracing a global legal framework grounded on essentialist claims of authenticity.

Ayurveda, a rich and widely used medical system in India, finds mention in the *Rig Veda*, but it drew from and grew through multiple sources and influences.[69] In the continued evolution of medical systems within India in the background of accelerated industrialization, intellectual property rights challenge the historical role of Ayurveda as a collective resource that should be protected from private appropriation, even while new, innovative products seek protection through trademarks and patents.[70]

Most importantly, with the rise of Hindu nationalism in the last few decades, the valorization of the indigenous grows even more political and problematic. The ancient Vedic sciences are routinely polished and presented as ancient wisdom and traditional knowledge.[71] Arun Agrawal calls such claimants the "neo-indigenistas." Ironically, as he argues, such groups often using the dictates of science and the digital to validate and celebrate the pure, authentic, and "indigenous."

> In their desire to find an elevated status for indigenous knowledge, they attempt to use the same instruments that western science uses. In so doing they undermine their own assertions about the separability of indigenous from western knowledge in three ways: 1) they want to isolate, document, and store knowledge that gains its vigour as a result of being integrally linked with the lives of indigenous peoples; 2) they wish to freeze in time and space a fundamentally dynamic entity—cultural knowledge; and 3) most damning, their archives and knowledge centres privilege the scientific investigator, the scientific community, science and bureaucratic procedures.[72]

Agrawal draws our attention to the insidious blurring of lines between representations of indigenous and western knowledge systems where it is those in power who get to articulate and make claims. In India, medicinal plants are mentioned in mythology. "Retro-botanizing," Projit Mukharji argues, reduces nonmodern plants to "local knowledge" to highlight their practical curative properties.[73] In a fascinating essay on retro-botanizing, he argues that "cultural pasts of a plant-object are not mere antiquarian curiosities, just as retro-botanizing is not a transparent and mechanical action. Retro-botanizing actively negotiates multiple cultural pasts of a plant and attempts to fix a stable identity."[74]

For the TKDL, culture must be frozen in space and time to package a knowledge claim. How do we move through these convoluted trajectories, a veritable maze, of knowledge about plants? In a sustained exploration of the TKDL in the context of Ayurveda, Gaudilliere argues, "value is not an ontological category, whether an ontology of reproduction or an ontology of labor, but the outcome of market-oriented processes, including processes of standardization and regulation."[75] These complex and historically located circuits profoundly shape plant histories. The cinchona tree, often a key example of the cultural appropriation of traditional knowledge, became modern and global (its bark is prized for quinine and other alkaloids, the only effective treatment of malaria during the height of European colonialism), while the neem plant, discussed earlier, became local and traditional.[76] These stories remind us how intimately plants and botanical knowledge are linked to histories of colonialism and its afterlives.

If not for postcolonial historians of science, the *Hortus* would singularly extoll colonial science. If not for the Ezhava and their claims, the Indian state would claim the *Hortus* as "Indian" (read Hindu and upper-caste) knowledge. The systematic colonial erasure of local experts, often along caste and class lines, points to the many strategic silences and gaps in the historical record.[77] Without historians of Ayurveda, its claims of an upper-caste and Vedic knowledge would remain unchallenged. A decolonizing project isn't about recovering an authentic, pure knowledge system from this thoroughly braided and impure history. Rather, it is to attend to the hegemonies of power.

THE MANY LIVES AND AFTERLIVES OF COLONIALISM

Colonial logics, regimes, and infrastructures live on in the postcolonial state. As early third world critics pointed out, in the name of modernity and development, science and technology grounded the state policies and priorities of independent India.[78] Postcolonialism is marked by embranglements, by hybridity, by cross-pollinations.[79] There is no site of purity to cling to, no possibility of return. Postcolonial theory has taught us that colonialism has changed both the colony and the colonial power forever. *Hortus* and TKDL show us the intimate embranglements of layered history in South Asia. Postcolonial elites, often British-educated, became the new ruling class. In newly independent India, science lived on as a central imperative, indeed as a "reason of state."[80] Western development projects that promoted industrialization and science and technology were prioritized. From the point of view of the subaltern—those who were not colonists or Indian elites—what did independence mean? Did the subaltern, those in the margins of power, truly become free or independent? Two examples illustrate the complexities of power in India, especially the politics of caste.

First, during the freedom movement against the British, as the blueprint for an independent India was crafted, caste loomed large. It was clear that most of the founders of India, members of upper castes, imagined an India that replicated the caste system. B. R. Ambedkar, a jurist who was one of the chief architects of the Indian Constitution and who was also a Dalit, looms large as a key figure. So disillusioned was he by the process and the final result that he converted to Buddhism and inspired a Dalit Buddhist movement, one that would refuse Hinduism and its casteism.[81] Ambedkar was not alone in his concern about caste. Some anticaste activists, especially in the state of Tamil Nadu, in fact opposed the departure of the British from the country without settling the issue of casteism. What was the difference, they asked? As revered anticaste activist Periyar (E. V. Ramaswamy) argued, independence meant a transfer of power from the "British to the Brahmins and Baniyas. . . . It was not independence in the true sense." Indeed, the political party Dravidar Kazhagam (DK) advocated that their followers boycott the Independence Day celebrations. K. Veeramani, president of DK, argued that Periyar believed more in social integration than in national integration, arguing, "He believed that if you are sincere about an Independent India, it must be independent of caste."[82] This is a profound statement about casteism—from the point of view of a Dalit, independence from the British was a renewed dependence, and perhaps a more oppressive dependence, on Indian elites. Given recent images of violence in contemporary India, this fear seems prescient for some and inevitable for others.

The second historical example is that of language in India. In consolidating a diverse conglomeration of territories across India into a nation, language was a primary challenge. There was no language that united everyone, not even a majority of Indians. The problem is that with more than 19,500 languages and dialects,[83] language has always been divisive in India. In an attempt to unify the nation, the founders of India settled on two official languages, Hindi and English, but no national language. Language politics simmer on today. Hindu nationalists have long posited that India should have one language, namely Hindi. In the words of the home minister Amit Shah, "Diversity of languages and dialects is the strength of our Nation. But there is a need for our Nation to have one language so that foreign languages don't find a place. This is why our freedom fighters envisioned Hindi as Raj Bhasha [the official language]."[84] Yet far less than half the country knows, let alone speaks, Hindi. Vast swaths of the country, especially in the south, west, and east, do not speak Hindi and have actively fought against its imposition. Indeed, English, the language of British colonial rulers, has emerged as a more unifying alternative. Why ever not? Indeed, English has evolved differently in different parts of the globe, emerging as a "glorious mess of a language."[85] English in India has its own unique syntax. In fact, colleges teach "Indian English" literature as a separate subject from English literature. In a meditation on language, Yogita Goyal writes that this is "a way to be a native reader but not a nativist, to treasure languages not in terms of ownership but of memory, not to exclude but to create community, and to imagine relation."[86] Indeed, Arundhati Roy notes that in 2011 Dalit scholar Chandra Bhan Prasad built a village temple to the Dalit goddess of English. "She is the symbol of Dalit Renaissance," he said. "We will use English to rise up the ladder and become free forever."[87] It is crucial to note that English here is not a tribute to empire but resistance to the history and local politics of caste in the subcontinent.

With these two examples firmly in mind, let us return to botany.

COLONIALITY OF NOMENCLATURE

Juxtaposing the *Hortus* with the TKDL is a powerful story of what I call the coloniality of nomenclature. Botany remains grounded in a powerful colonial script—a goal toward a systematized botany that runs through projects such as these. These colonial scripts and lexica shape modern regimes of science and law. Once independent, postcolonial nations continue to be wrapped within the luxurious folds of the norms of colonialism. We see the coloniality of nomenclature playing out in many layers.

First, as Van Rheede argued, creating the *Hortus* was necessary to aid the extractive goals of empire. The names of plants were eventually translated into Latin. Some were named after famous white explorers, and some had Malayalam names latinized into a binomial system. In the TKDL we see how third-world countries resisting western patterns are still caught up in the same colonial scripts, quite literally. In order to preserve long-enduring and well-established knowledge practices, they need to prove "prior use." Again, the west remains at the center of power, while others are forced to adapt to its normative language and legal systems. Colonial appropriation follows into neocolonial appropriation. Coloniality of nomenclature captures the institutionalization of colonialism in the infrastructures of global regimes of commerce and their knowledge infrastructures, including botany. Colonialism is clearly not "post" but endures on.

Second, there is no site of purity and no easy return. In the case of India, there was no India before 1947: it was a colonial invention as South Asia was carved into the two nation-states of India and Pakistan during the Partition. Further, cases like the *Hortus* teach us an important lesson—that we cannot entirely reject colonial knowledge because it includes the (unevenly) appropriated knowledge systems of the colonies and of the subaltern. In some sense, colonial knowledge is itself hybrid.

Third, elite members of the postcolony proved to be no saviors of the nation, often enriching and empowering themselves instead of the nation. They enforced regimes that strengthened their hold on the nation—using the state apparatus to continue the privileges of gender, caste, class, ability, and religion that favored them. As members of the Ezhava community rightfully claim, the *Hortus* belongs not to India but to them.

Fourth, contemporary politics of Hindu nationalism, while ostensibly claiming to be both anticolonial and anti–postcolonial elite, in fact reinforces both rather than rejecting the colonial or postcolonial script. By a facile politics of embracing western science as Hindu science, Hindu nationalism extends the western colonial project within its folds. Rather than take on the complex and long history of caste, gender, and class, it refuses to truly engage with the heart of anticolonialism in the name of an imagined Hindu nation.

Finally, one issue I have not explored, because of a scant literature, is the complex history of botanical nomenclature. While much of botanical nomenclature is latinized and draws on European names, this is not true in all cases.

In a fascinating history of the Indian spikenard, Minakshi Menon shows how William Jones encoded native experience and the plant's known therapeutic value to create a botanical classification that forced the plant "to travel across epistemologies and manifest itself as an object of colonial natural history."[88] Other examples where local names produce more reliable and stable nomenclature challenge any easy story of European epistemologies.[89] Menon urges historians to uncover the multiple and rich traditions and tensions within the history of botanical nomenclatures.

CONCLUSION

In analyzing the long arc of colonialism and postcolonialism in India, we need to move beyond the binaries *colonizer* and *colonized*. The British lay the infrastructures of colonialism, and while some of the colonized resisted, some upper-caste groups like Brahmins aided and abetted colonialism. While some colonizers recognized the importance of the local, others genuinely believed in the epistemological superiority of the (European) enlightenment.[90] Any project of decolonization needs to reckon with these complexities.

What do these two cases that span four centuries tell us? In the context of the coloniality of language, Charu Singh reminds us that "Far from being a matter of linguistic translation alone, the construction of word-level equivalence required linguistic, epistemic and political strategies to render nomenclature meaningful, stable and authoritative for its vernacular publics."[91] The coloniality of botanical language and nomenclature created deep, enduring, and layered infrastructures. Neocolonial processes of trade have only consolidated these circuits.[92] As we undo one layer of the infrastructure, new ones emerge.

Much as we herald the power of globalization in the contemporary world, globalization has a much older history. Plants have been at the center of transnational circuits of trade for centuries and have grounded the extractive logics of biopiracy ever since. Colonial plants continue to adorn western gardens, herbaria, and museums. Contemporary plant sciences continue to be centered in the west. If western scientists study the biodiversity of the colonized world, they "parachute" into these nations to gather the material they need to bring them back to the west.[93]

And yet any understanding of postcolonial nations reminds us that west/

east is too facile a binary. Postcolonial elites have established their own hierarchies of power in the postcolony.[94] From the point of view of marginalized populations in India, the exploitative power of both the British and those of the upper-caste Brahmins remain powerful and oppressive. If trade is not fair, it can never be free.

We have moved from the kitchen table to the world. It is the kitchen table—the cultural politics of food—that fueled so much of colonial conquest. Within South Asian medical systems and local practice, food, health, and medicine are closely intertwined. The powerful work of the Kitchen Table Press over thirty years ago pressed the point that none of us can be free until we are all free.[95] In discussing the site of the kitchen table in his recent book, Cameron Awkward-Rich offers us a trenchant reminder of how histories of exclusion persist in trans studies, as queer and trans communities of color are often erased in contemporary trans studies.[96] Colonialism lies deep in intellectual and academic infrastructures. Thus, even newer fields like feminist, ethnic, queer, trans, and disability studies replicate these exclusions. Just as colonization sedimented its ambitions through epistemic infrastructures, decolonization is a project that is even more complex as we confront the layered hierarchical infrastructures of names, norms, and nomenclatures.

Fables for the Mis-Anthropocene: Making a Little Trouble Everywhere

Noor was excited. Two weeks ago, she had uploaded the pictures and sounds of worms on the mulberry tree. They had caused quite the sensation. Scientists in a nearby field station had written, wanting to see them. Noor agreed. Salma walked with Noor and the scientists to the nearby garden. Researchers took samples of the worms and the mulberry leaves. The head of the station called Noor a few days later just before a press conference. This was a new species of silkworms! They named it after the station head, *Bombyx wilsoni*, and announced, "We believe this is a domesticated silkworm that escaped cultivation and has adapted to a new habitat. The mulberry trees also seemed different, producing a new chemical to ward off predation. But the silkworm has adapted to the new chemical by learning to chew the leaves so as to not release the poison." *That* was the different sound Noor had heard.

Over the weekend, Noor went back to the gardens with her friends Samira, Kyle, Nina, and Jaylin to show them the trees and worms. As she told them the story, they were outraged. "They should name the new species after you!" said Nina.

"No!" cried Noor. "I agree I am more deserving than that head, but worms should not be named after human beings. I didn't discover anything. The worms knew who they were!" Everyone burst out laughing. "Worms and plants should not have human names. We should describe them as they are and give them names that reveal their true selves. I would have called them the wide-jawed silkworm." Everyone nodded in agreement. Noor made a good point. "Let creatures be themselves," they all cried.

As the children looked closely, the mulberry leaves on this stand of trees were also different. They had tiny specks that other trees did not. Why was the mulberry tree also not a new species, they asked Noor.

"Right?" Noor exclaimed. "I asked them exactly that. 'Well,' the station head told me, 'the trees are genetically too similar to the other mulberry trees. You see," he said, "laypeople cannot tell the true identity of species. We scientists extracted the DNA from the worms and the trees and looked at their genetic codes. That small difference you saw was not trivial genetically. These worms can now only live on these trees. Only such meaningful change can create a new species. While earlier, taxonomists classified organisms by how they looked, their morphology, it was not very accurate. After all, bats and pigeons fly, but bats are mammals, not birds. Evolution produces all kinds of variation, but the essence of an organisms is its DNA; only that can tell us its true identity. So the true classification system is a classification based on DNA.'"

Noor and her friends pondered. "So an organism's DNA might have changed a lot, and it could be called a new species even when it did not look so different? But then, we cannot name anything accurately! We would always need to call the scientists to come and grind up the plant. Not fair, they want to control everything!"

"Why does there need to be only one 'truth'?" asked Samira. "I think evolution is cool, and the scientists are showing us something important. If our purpose is observation and appreciation, why can't we use our senses to name the world around us? Scientists constantly talk about wanting more diversity yet do not include us and the diversity of our thinking. Scientists can have their traditions, why impose them on us?"

"Scientific names are mostly Latin and Greek and named after colonial explorers," cried Jaylin, outraged. "I want no part of that. We should reclaim our world. We should get to name them. There are so many problematic and racist words we do not want to use."

"What is important," Kyle suggested, "are not only individuals but communities, how organisms live together. Think of the coevolved strawflower and its only pollinator, the straw moth—the flower looks like a straw, as does the moth's proboscis. Surely, they should share a name?"

The Chirp-Net was abuzz over the months following. They concluded that they needed good morphological data—good descriptions and pictures. New standards. Children agreed on how best to take a picture, how to standardize size, which parts and angle of the plant were important. The web erupted with funny names, clever names, silly names, serious names. Children from across the world renamed plants with wild abandon.

"Taxonomic anarchy!" scientists cried. But the children were resolute. "You learn our languages," they said. Of course, thanks to the multilingual catalog of Chirp-Net, it took just a button to figure out the many names of the organism. Scientists could also figure out what they wanted to. There was no reason to impose their system on everyone.

And so the children remade the world, creating new communities of naming and belonging. These worlds were within their homes, gardens, woods, forests, skies, streams, rivers, and oceans—worlds teeming with loving ecologies. By making a little trouble everywhere, this is how the Raucous Spring remade the world into one filled with curiosity, play, and joy, a world always in the making, being unmade and remade continually.

INTERLUDE

An Ordinary Botany: Haunted Archives of Livingness

As a child of Bollywood, I always found ghosts and hauntings to be moments of reckoning for me. Some years ago, ghosts harmed by eugenics animated my book *Ghost Stories for Darwin*. Ghosts were, for me, abject figures. In listening to the dispossessed, I reckoned with my biological education and strove to think anew about biology's eugenic pasts. Katherine McKittrick in *Dear Science* presents a very different genealogy of hauntings and ghosts: not abject ghosts of the haunted archive but lively ones filled with the relish, wonder, creativity, beauty, and joy of black livingness and relations. She opens a door into new, generative vistas of the past in a way that made me rethink my own botanical genealogy.

In India, one must begin with food. Much of my childhood was spent climbing trees to find the tart mango, the sour gooseberry, the green guava, the tangy tamarind pod. With a little salt and chili powder, it was transformed. Indeed, spices were the key to life. Every dish had a different combination of spices, each one distinct: anise, asafetida, bay leaf, black cardamom, black cumin, black pepper, carrom, cayenne, cinnamon, cloves, coriander, cumin, curry leaves, fennel, fenugreek, garlic, ginger, green cardamom, mace, mango powder, mustard, nutmeg, onion seeds, poppy, saffron, turmeric, white pepper, and a variety of chilies each with its own special flavor and purpose . . . a basic repertoire for any serious Indian cook. You learn early in life to sniff the wafting smells to discern the combination of spices and guess what your neighbors were cooking. You then decided which friend you wanted to visit that day! I'm sorry, but salt, pepper, and the often recommended spice mix DASH don't cut it for me. It seems an impoverished life. I could tell you tales about friendship, play, music, movies, summer holidays, religiosity, rituals, and mythology—all of which filled my childhood with botanical worlds.

Plants were central to play. My friends and I would go in search of a solid stem of what we called friendship grass. We delicately bisected both ends of the stem and gently pulled it apart. As the two ends met in the center, they would not split up but rather formed a square. We would exchange one end and then pull it slowly apart as the square grew, bigger and bigger. The larger the square, the deeper the friendship. It was silly, I know. But good friends roamed many a day among the roadside grasses to find the perfect specimen, the right texture of stem, fibrous but strong. Then there was the carrom plant. You could flick the top of it with your finger just like a carrom coin. Who could flick one the farthest? We honed our skills to evaluate the inflorescence—its size, the strength and turgidity of the stem—to determine how far flower head would travel.

Indian music is all about plants. Jackfruit was delicious, but the veena owed its deep tones to jackfruit wood. The tambura, mridangam, violin, and *dhol* are made from woods of trees. A dried gourd shell can transform simple notes into resonant tunes. The flute emerged thanks to bamboo. Reeds give sounds to the nadaswaram, the shehnai, the recorder, and the saxophone. The hides of cows, sheep, and iguanas, stretched across wood, generate the inimitable timber of the tabla, mridangam, or kanjira. One of the joys of childhood was finding various reed grasses and blowing through them to produce distinct sounds. You could find music in unexpected places in the concrete jungle.

When I was in middle school, we moved cities and lived across a famous park in Chennai called Nageshwara Rao Park. The park had beautiful trees, especially some memorable fan palms. We ran and walked through the winding paths, green grass, and lush gardens. A well-maintained park, it was a favorite location for movie sets. We would crowd around and spend many hours watching famous actors ridiculously prance around the garden. Indian movie songs are filled with backdrops of botanical beauty.

The tulsi plant is often the center of the Hindu home. The basil-infused water of the temple is distinct and refreshing. Mango leaves and marigold flowers grace an entrance. The betal leaf and turmeric root accompany gifts. Mythology is filled with plant life—the antidote from the sanjeevani tree that saves Lakshman, *Shabari* and her humble fruit offerings, *Sudama* and his flattened rice. Popular stories like the one of Vikram and Vetal infused popular trees like the peepal with deep histories.

When I entered graduate school in the United States in the biological sciences, I quickly realized that my upbringing and relationship to plants were most odd. The concrete jungle was not nature and was seen as outside the purview of the pure biological sciences. My peers, on the other hand, were knowledgeable about the nature we were supposed to care about. They grew up hiking in old-growth forests. They had hiked the Appalachian Trail, backpacked across Europe, canoed and kayaked through marine ecosystems. They had camped night after night, soaking in the sounds of the wilderness, a world removed from humans. They had faced curious coyotes and angry bears. They talked the language of biodiversity, botanical nomenclatures, herbarium sheets, and conservation biology. They knew firsthand what we read in our textbooks—plant succession, competitive exclusion, mutualisms, changing ecologies—and had directly experienced the wealth of diverse ecosystems. Faced with this breadth of knowledge, my ignorance was sobering.

For my part, I had faced the Indian flying cockroach and the house lizard, and its twitching tail left behind to confuse us; likewise, my love of plants was entangled in a botanical knowledge of the everyday. It was a passionate relationship. It was built around not the wilderness but the plants of the gardens, the roadside, the kitchen, and the lush growth in the cracks of the concrete, and in my urban imagination that marinated in books and the gorgeous vistas from television. During my entire academic career, I have seen my love of plants as suspect. I understood them to be entangled and intertwined in the many unnatural worlds of urban ecologies and popular films—worlds seemingly removed from nature and the natural. And I have been ashamed. Ignorance translates into inferiority, and its stench stays with you. It is only through the work for this book that I began to understand how colonial histories have shaped botanical foundations. I have come to appreciate the botanical worlds I grew up in. I loved them and they inspired me, and this is a legitimate genealogy. So much of the botany of my textbooks—botanical nomenclature, ecological theories, conservation biology, and biodiversity—are deeply rooted in colonial logics.

In *Dear Science*, Katherine McKittrick's evocative prose and poetry open up black life and black livingness as inventive and rebellious methodologies that should be taken seriously.[1] Who, as McKittrick argues, is to say what legitimate knowledge is? I too should claim my botanical heritage as proper, as natural, as normal. I too am a lover of plants, albeit those of the concrete jungle. It may not be classical botany, but it is a botany—an ordinary botany, and I, an ordinary botanist.

PART THREE

Floral Dreams

SEXING REPRODUCTIVE BIOLOGY

Throw massive amounts of water and petrochemicals
on your grassy plot, let it push up from the soil,
then cut it down to nubs before it can grow up
and have sex and go to seed. A lawn is an
endless cycle of doomed ecology.

ELLEN MELOY | *Seasons*

CHAPTER FIVE

The Orchid's Wet Dream

Sex Told, Untold, Retold

> Biological Exuberance is, above all, an affirmation of life's vitality and infinite possibilities: a worldview that is once primordial and futuristic, in which gender is kaleidoscopic, sexualities are multiple, and the categories of male and female are fluid and transmutable. A world, in short, exactly like the one we inhabit.
>
> BRUCE BAGEMIHL | *Biological Exuberance*

> Oh to be a pear tree—any tree in bloom!
> With kissing bees singing of the beginning of the world.
>
> ZORA NEALE HURSTON | *Their Eyes Were Watching God*

WHAT DO FLOWERS DREAM OF? Years of botanical education evoke a florid scene: vibrant colors, sensuous petals, redolent sweetness, languid lines, viscid stigma, frisky pollen, imaginative kinesics, grand seductions—a dynamic, sumptuous sensory feast of sight, sound, smell, touch, and taste. The evocative, flowery, and flamboyant tales of plants impassion budding biologists. A botanical education is suffused with stories of sex that are at times eerily reminiscent of humans. Indeed, stories we tell about plants, plant parts, and plant sex inevitably rationalize overdetermined racialized cis-heterosexual scripts. For those not in gender studies this phrase, "overdetermined racialized cis-heterosexual scripts," may seem like a mouthful. But it synthesizes decades of feminist scholarship. It refers to how the binaries of cis-bodied males and females are forever woven into heterosexual scripts of (raced) reproduction. It is overdetermined because we assume it even if it isn't there. For those of us who do not fall into the categories of the normal, it is difficult to overstate how all-encompassing and hegemonic these structures and scripts are. This chapter explores the history and ideological grounding of these scripts. Reading scientific literature as "text" has a long tradition in feminist STS.[1] It brings together the humanities and the sciences to illuminate how language, metaphors, and cultural tropes mediate and make meaning of our worlds.

The stories we tell about plant sex share deep histories with much of biological storytelling. In their famous critique familiar to most biologists, Stephen Jay Gould and Richard Lewontin warned us about adaptationism or the "adaptationist programme." In their influential paper, they warn against biological stories that rationalize various characteristics as adaptations.[2] These stories typify what they call the Panglossian Paradigm, invoking Voltaire's ridicule of Dr. Pangloss when he claims: "Things cannot be other than they are. . . . Everything is made for the best purpose. Our noses were made to carry spectacles, so we have spectacles. Legs were clearly intended for breeches, and we wear them."[3] Their critique led to the term *"just so" stories*, cautioning us to not assume that every characteristic that exists in the world has evolved because it is adaptive. After all, nonselective forces such as phyletic or developmental constraints, randomness, and pure chance can shape evolution. For example, rather than adaptation, exaptation (biological traits developed for one function remodeled for a different use) may equally explain various traits and biological characteristics.[4] In short, we need to be careful about making up "just so" stories to explain the evolution of life.

Like most other fields in biology, plant reproductive biology is suffused with "just so" stories—stories about plants, plant parts, and plant sex that inevitably rationalize and naturalize overdetermined racialized cis-heterosexual scripts. Here I attempt a queer reading of botany's account of plant reproductive biology. I tell the story by denaturalizing the scripts and making us see them in a new light. In three acts, I read, unread, and reread plant reproductive biology to ask how we might dismantle regimes of racialized cis-hetero patriarchy.

Rememorying a Queer Planet

In rehistoricizing the history of botany, feminism, and the planet, I return to Toni Morrison's evocative concept "rememory" to retell floral biologies' embrangled histories. Plants challenge the tired old language of empire. They challenge us to avert the gaze of androcentrism and Eurocentrism to imagine plant worlds anew. Plants urge us into new conversations to reimagine what we mean by sex and sexuality.

ACT I: LINNAEAN LIBERTINES, OR WHY PLANTS HAVE SEX

In any analysis of colonialism, sexuality emerges as a critical node in plant biology. Sex is a surprisingly capacious term that means many things. I focus on two aspects, intimately related and critical to biology—sex as an act of intercourse/copulation, and sex as a purported state of the body. Sex and gender are intimately linked—sex is always gendered. Sex and sexuality as we understand it today is a recent invention, emerging only about 150 years ago.[5] The extent of its naturalization and normalization is a testament to the power of colonialism. In evolutionary biology, the emergence of sexual reproduction is heralded as a critical innovation that produces variation, the terrain and playground for natural selection. In this chapter, I focus on angiosperms, or flowering plants, about which theories of sex and sexuality are most evocative of theories of animal and human sexuality. Rememorying the planet forces us to contend not only with the embrangled worlds of human, plant, and animal biologies but also with how scientific knowledge was founded on the twinned binaries of sexual difference and racial difference. The two are never separate.

THE SEXUAL LIVES OF PLANTS Londa Schiebinger's *Nature's Body* remains an important and indeed a foundational analysis of the emergence of plant sexuality, chronicling the legacies of Linnaeus.[6] My account draws heavily from her work. She opens the book with the following:

> Hermaphroditic plants "castrated" by unnatural mothers. Trees and shrubs clothed in "wedding gowns." Flowers spread as "nuptial beds"

for a verdant groom and his cherished bride. Are these the memoirs of an eighteenth-century academy of science, or tales from the boudoir?[7]

As Schiebinger reveals elsewhere, Carl Linnaeus developed new sexual vocabularies to revolutionize the study of the plant kingdom.[8] Prior to Linnaeus, within medieval cosmology, plants were understood for their primary purpose: food and medicines. As late as the Renaissance, botanists named parts we consider sexual today with nonsexual vocabularies—the stamen from the Latin term denoting the warp thread of fabric, the pistil or pisl because it resembled a pestle.

Most historians have argued that Linnaeus's vision was shaped by normative ideas of sexuality premised on binary sex and cis-heterosexuality. Human sexuality was naturalized and universalized through its imposition onto the worlds of plant sexuality. Schiebinger argues that in the seventeenth and eighteenth centuries, precisely in the times of Linnaeus, naturalists became interested in assigning sex to plants.[9] The description by Nehemiah Grew, who first identified the stamen as male, is striking:

> The blade (or stamen) does not unaptly resemble a small penis, with the sheath upon it, as its praeputium [prepuce]. And the . . . several thecae, are like so many little testicles. And the globulets [pollen] and other small particles upon the blade or penis . . . are as the vegetable sperme. Which as soon as the penis is erected, falls down upon the seed-case or womb, and so touches it with a prolific virtue.[10]

Grew thus ascribed maleness to the plant because the stamen looked and functioned as a penis did in animals, declaring that every plant is . . . male or female.[11] By the early eighteenth century, the analogy between animal and plant sexuality was fully developed, and most biologists agreed that plants had sexes and used them to reproduce.[12] However, most work focused only on male parts, and exclusively on sexual reproduction and the heterosexuality of plants. Linnaeus considered stamens and pistils to be "the very essence of the flower," bringing traditional notions of gender hierarchies "whole cloth into science" and incorporating fundamental aspects of human social order onto the natural worlds.[13] He blurred the differences between plants and humans[14] and developed elaborate analogies of convergence: filaments of stamens were the vas deferens, anthers the testes; pollen was the seminal fluid. In the female, the stigma was the vulva, the style the vagina, and the pollen tube the fallopian tube; the pericarp was the impregnated ovary and the seeds the eggs. Others even argued that the nectar in plants was equivalent to mothers' milk in humans.[15] Plants that reproduced through vegetative and clonal means were deemed *asexual*, and those with spores (without flowers or seeds), such as ferns, mosses, algae, and fungi, as *Cryptogamia* (plants that marry secretly).

Linnaeus gave primacy to plant sexuality, and his "scientization" of botany coincided with an ardent sexualization of plants. Schiebinger argues that the scientific revolution and the revolution in sexuality and gender came together to elevate plant sexuality as a central focus of botany. Janet Browne concurs, arguing that "to be a Linnaean taxonomist was to believe in the sex life of flowers."[16] Linnaeus based his system on sexual difference, and the "laws of nature" were read through the evolving lens of human "sexual relations." As Schiebinger argues, Linnaeus based his system on sexual difference, giving male parts priority over female parts in determining the status of organisms in their larger classificatory structures. This male-dominated world sat squarely on a binary heterosexual worldview: "Not only were his plants sexed, but they actually became human," specifically husbands and wives.[17] His renowned "Key to the Sexual System" is founded on the *nuptaiae plantarum* (the marriage of plants)—based on their union and implicit heterosexuality, not their sexual dimorphism.

Linnaean classification was based on a sexual morphology and claims of difference between the male and female parts of flowers.[18] In his nomenclature, he did not use the nonsexual terminology of pistil and stamen but introduced *andria* and *gynia*, which are derived from Greek for husband (*aner*) and wife (*gyne*). Flowers that lacked stamens or anthers were termed *eunuchs*, and the removal of anthers was called *castration*.[19] He divided the "vegetable world" into classes based on the number, position, and proportion of the male parts, or stamen. These classes based on male anatomy were then subdivided into 65 orders based on the number, position, and proportion of the female parts, or pistils. These were then divided into genera based on other flower parts, like the calyx and corolla, then into species based on leaves and other characteristics, and finally into varieties.

The terminology was not only sexual but grounded in patriarchal and Eurocentric ideals of family structures. Plants gave rise to a botanical landscape onto which naturalists projected western ideals of human sexuality. Indeed, the Linnaeus of *Nature's Body* is not a dispassionate observer of nature but rather a naturalist who actively shaped his observations to conform to a paternalistic and patriarchal society that celebrated the virtues of female domesticity.[20] His new terminology gave rise to a classification system where classes of plants end in *andria* (*monandria*, *diandria*, and so on) and orders in *gynia* (*monogynia*, *digynia*, etc.). Note, class/male is ranked higher than order/female. In short, as Schiebinger argues, "Linnaeus saw plants as having sex, in the fullest sense of the term."[21] He borrowed terms for romance, eroticism, and the sanctity of love, nuptials, and marriage into his vocabulary for plants. Plant sexuality took place almost exclusively within the bonds of marriage. Through Linnaeus's imaginative classification system, lowly plants were transformed into humans, and their sexual morphology and behavior came to resemble human sociology and sexuality.

Linnaeus's impact cannot be overstated. As a colleague of his quipped, "God created and Linnaeus organized."[22] What made the Linnaean system so popular was its simplicity. As biologist Ernst Mayr explains, "Any botanist using the sexual system would come to the same result as Linnaeus. All he had to do was to learn a rather limited number of names of the parts of the flower and fruit and then he could identify any plant. No wonder nearly everybody adopted the Linnaean system."[23] Linnean taxonomy, built from new understandings of human sexual difference and traditional notions of colonial sexual hierarchy, was imbued with the vivid sexual language of humans. The importance attached to sex by both the church and the state, and their regulation of sexuality, shaped the growing importance of its biopolitical dimensions.[24] The implicit use of gender to structure botanical taxonomy came together with the explicit use of human sexual metaphors to create a new and innovative classification system for plants.

This innovation arrived during the seventeenth century, at a time during the Age of Exploration when academic botanists began to break ties with medical practitioners. Exploits from colonial voyages of discovery and new plant materials from the new European colonies flooded Europe; the number of known plants quadrupled between 1550 and 1700. The proliferation of materials and knowledge opened up new methods of organization as botanists sought simple principles that would hold universally. By the end of the eighteenth century, the Linnaean system had beaten the others and emerged as the norm.[25] While it has subsequently given way to more complex taxonomies, much of its terminology and the approach of binomial nomenclature persist today. In fact, the *International Code of Botanical Nomenclature*, which started in 1905, continues to regard the *Species Plantarum* (1753) and the fifth edition of *Genera Plantarum* (1754) as the official starting point for botanical nomenclature.[26]

Linnaeus's legacy goes beyond plants. He also classified humans to fortify the precision of race by ordering his four "races" into a hierarchy.[27] Linnaean racial typography has also had a far-reaching legacy. As feminists have long reminded us, gender is a racialized category. In this colonial vision of evolution, Europeans remain at the apex to produce the most manly men and womanly women. Non-European men were rendered effeminate and womanish (degenerate men), and non-European women were considered more manly, deformed and not feminine.[28] This enduring investment in a shared model of classification remains Linnaeus's legacy.

The eighteenth century saw botanists' classification of plants shift from a more local form of "useful" and hands-on knowledge toward one of scientific classification that was at once both abstract and universal. Linnaean nomenclature was precise and revolutionary, bringing the abstract precision of mathematics into biology, alongside what Erasmus Darwin called the "loves of plants."[29] Botany's colonial legacies are immense and deep, as Linnaean taxonomy traveled through the colonies. As Browne argues, "Just as the British Museum and Kew Gardens were constituted by the flora, fauna, and human knowledges extracted from the colonies, the discourse of natural history was articulated in terms of biotic nations, kingdoms, and colonists, reflecting the 'language of expansionist power.'"[30] Classification and nomenclatures are deeply rooted in the politics of their times, as colonial expansion permeated the globe and western systems of nomenclature supplanted local knowledges. Around the globe, colonial legacies dismantled and erased a plethora of languages, meaning-making practices, and nomenclatures to usurp flourishing cultures of knowing with universal, scientific "monocultures" of knowledge.[31] Antiguan writer Jamaica Kincaid evocatively notes:

> Who has an interest in an objective standard? Who needs one? It makes me ask again, What to call the thing that happened to me and all who look like me? Should I call it history? And if so, what should history mean to someone who looks like me? Should it be an idea; should it be an open wound, each breath I take in and expel healing and opening the wound again, over and over, or is it a long moment that begins anew each day since 1492? . . . This naming of things is so crucial to possession—a spiritual padlock with the key thrown irretrievably away—that it is a murder, an erasing.[32]

Through Linnaeus, plants inherited human sexual and social vocabularies, which were in turn grounded in ideas of Christianity. Mary Louise Pratt argues that Linnaeus ushered in not just a global but a larger "planetary consciousness" that shaped European colonialism and exported a "natural" order (shaped by Christian morality) under the guise of science.[33] Liberal thinkers put Christian ideas about "natural" and unnatural" sex on a "scientific footing." Social reformers in turn invoked this scientific nature to reorder and reorganize gender, sexuality, and family life across colonial states.

BOTANY AND THE LEGACY OF LINNAEUS The study, collection, and use of plants is a particularly gendered affair—by the eighteenth century, botany was one of the few scientific disciplines open to women, and the century saw many prominent female botanical collectors, illustrators, and experts.[34] The sexualizing of botany and the plant world had a profound impact on women who were until then avid botanizers.[35] This was contentious, complex territory. While Victorian sexual ideology saw women as passionless, women's engagement with botany was also an engagement with God's work.[36] The Linnaean classification system grounded in a "marriage of plants" imagined households of multiple wives and husbands. During a period that saw a rise in modern pornography, this vision scandalized and stifled, and botany, once a women's science, shifted out of women's reach. In Linnaeus's descriptions, only one class, Monandria, practiced monogamy. In contrast, the class Polygamia lived with wives and lovers (harlots and concubines) in marriage beds.[37] There is no evidence that Linnaeus, the son of a Swedish country parson, intended the pornographic undertones, but they still scandalized. German scientist Johann George Siegesbeck argued, "chaste young people should be protected against Linnaeus' sexual gospel, which harboured lewd features. . . . Who could ever imagine that Almighty God would establish such a muddle—or rather such shameful whore-mongering—in order for plants to reproduce," and "who could teach young students such a lecherous system without causing offence."[38] Richard Polwhele's poem "The Unsex'd Females" (1798) entered public debates by lampooning women in botany and, more generally, coeducation, betraying a deep animus toward the women's movement and feminist activists such as Mary Wollstonecraft.[39] Indeed, this new body of work produced a "moral backlash" against females involved in botany—botanists, naturalists, artists, and literary writers.[40] The stereotype of the forward and sexually precocious female botanist emerged, illustrating both the contemporary appeal of the Linnaean system to women and the simultaneous anxieties surrounding female modesty.[41]

Linnaean history and its Eurocentric and racialized sexual mores and ideas of binary sex continue to ground plant biology today. Linnaeus's spirit of systematic categories stressing binomial nomenclature and the concept of species endures.[42] Today, much of biology, including plant taxonomy, has turned molecular, and has therefore moved from a frame of shared morphology to shared genetic sequences—from form to essence.

A quick look at plant reproductive biology easily demonstrates the impoverished frameworks of the European household. In comparison, there is breathtaking array of reproductive arrangements in plants. I began counting, but they proved too many.

Today, plants continue to be categorized by the "sex" of the organisms (a dubious and unstable concept). If a species has two sexes—that is, male and female plants—it exhibits *dioecy*, or two houses. Species are *monoecious* if they have only one house.

In monoecious plants, if stamens (defined as male reproductive organs) and carpels (defined as female reproductive organs) are found in different flowers, these are *male* and *female* flowers. If both are in one flower, they are *hermaphrodites*. It is important to note that over 85 percent of plant species are hermaphroditic, yet plant reproductive biology relies obsessively in the overdetermined categories of binary sex.

But hermaphroditic flowers do not all act the same, so we need new terminology. For example, male and female parts may mature at different times (*dichogamy*). If pollen function precedes, the plant exhibits *protandry*; if ovule, *protogyny*.

The male part is called a *stamen* and is composed of a filament and anthers producing pollen.

The female part is called a *pistil* and is composed of one or more carpels, each of which generally contains a style, stigma, and ovules.

These terms endure from Linnaeus, who created them as a version of human sexuality. For Linnaeus, these were literal analogies of convergence: filaments of stamens were the vas deferens; anthers, the testes; pollen, the seminal fluid. In the female, the stigma was the vulva; the style, the vagina; the pollen tube, the fallopian tube; the pericarp, the impregnated ovary; and the seeds, the eggs.

But in plants it is never so simple—nor in humans, for that matter. There are more exceptions than rules in the splendid exuberance of plant reproductive biology. Wikipedia offers a brief sampling of terminology:

androecious: having only male flowers
gynoecious: having only female flowers
androdioecious: having male flowers on some plants, bisexual ones on others
gynodioecious: having hermaphrodite flowers and female flowers on separate plants
androgynomonoecious: having male, female, and bisexual flowers on the same plant
andromonoecious: having both bisexual and male flowers on the same plant
gynomonoecious: having both bisexual and female flowers on the same plant
polygamodioecious: mostly dioecious, but with either a few flowers of the opposite sex or a few bisexual flowers on the same plant
polygamomonoecious: mostly monoecious, but also partly polygamous
subdioecious: having some individuals in otherwise dioecious populations with flowers that are not clearly male or female
subgynoecious: having mostly female flowers, with a few male or bisexual flowers

trimonoecious: having male, female, and bisexual flowers on the same plant

trioecious: having male, female, and bisexual flowers on different plants[43]

I think you get the picture! The proliferation of terms serves to accommodate a world of abundant and imaginative plant variation into the limited vocabulary of European sex lives.

You cannot but ask, "Why can't we count past two?"[44] Why look through the eyes of European puritanical sexuality to see only a binary cis-heterosexual male/female plant world? What if we do the reverse? How might the exuberant sexuality of plants allow us to recognize the exuberance of human sexuality? As Maja Bondestam provocatively argues, might we see Linnaeus not as a figure of conservatism and conformity of binary thinking, but as one who offered us "a full repertoire of nonnormative sexual combinations of stamens and pistils"?[45] This follows a recent trend to recover a queer Linnaeus, just as we have of Darwin.[46] Given the careers of Darwin and Linnaeus and the misogynist and racist politics of their legacies, however, I am reluctant to recover their legacy as feminist.[47]

Act II: An Orchid's Wet Dream, or How Plants Have Sex

Flowering plants, the most ubiquitous and diverse of groups, shape stories of plant sex. Let's begin with the usual story told about how flowering plants have sex. These popular stories highlight the promiscuous perversity of flowers. Garden columnist and horticulturist Bill Finch humorously captures the story in lively detail:

> Judging by the recent spate of garden questions I've received from reproductive-age humans, I'm tempted to believe there's an alarming and dangerous ignorance about the most fundamental facts of life. . . . So [I'm telling you what] apparently your parents were too embarrassed or too neglectful to tell you:
>
> Plants have sex. ALL plants do, or at least try to.
>
> And please forgive me for the graphic and coarse nature of the following content, but I must tell you that plants produce beautiful, elegant,

> fragrant flowers for one reason only: To have sex. And the reason they produce flowers to have sex is so that they can make babies. These babies (call them seeds if you like) are incubated in a womb that is more precisely known as a fruit. . . .
>
> Every plant you encounter—trees, shrubs, weeds, grasses—is at some point in its existence going to engage in this kind of scandalous behavior. Shocking, isn't it?[48]

As you can see, Linnaeus's anthropomorphizing endures to remake unfamiliar plants into familiar humans, and in so doing it naturalizes sex as the *raison d'être* of evolution, indeed of all life. If Linnaeus cemented the vocabulary of the plant sexuality in the familiar binary language of male and female and heterosexual coupling, then plant behaviorists translate human sexual vocabularies into plant behavior. But plants are not humans, and even humans are not this simple. Not by a long shot.

In a recent paper, "Re-imagining Reproduction: The Queer Possibilities of Plants," Madelaine Bartlett and I critique this popular story.[49] The above description by Finch significantly simplifies and mischaracterizes plant reproductive biology. Bartlett and I suggest new terminology and vocabularies that better capture plant biology. So much of plant biology emerged in the shadow of animal biology with little regard for whether the two are comparable.[50] As we argue, there are many differences between plant and human reproduction:

- Fertilization in human and plants are not the same, equivalent, or homologous. In animals, including humans, that are diploid (two sets of chromosomes), a reduction division, or meiosis, produces the haploid sperm and egg. In plants, however, the gametophyte is already haploid, so the sperm and egg are produced by mitosis (duplication rather than reduction division). Then, in an act of double fertilization, one gamete fuses with the ovule to form the zygote, and the second fuses with the two polar nuclei to form the endosperm that nourishes the growing embryo.
- Most flowering plants (unlike humans) can also propagate through asexual or vegetative means such as through roots, stems, leaves, and buds.
- The familiar vocabulary of male/female as universal, natural, biological, and essential is a fabrication. It is not an empirical observation

but a definition: In general, the individual that produces the relatively smaller gamete is universally called the "male" and one that produces the larger is called the "female."[51] The same definitions apply to the "female" and "male" functions in hermaphrodites.[52] This differential gamete size, or *anisogamy*, is at the center of our "just so" storytelling. Early on, the evolution of anisogamy took on an adaptationist story of a primeval conflict of the sexes.[53] For example, G. C. Williams argues that the evolution of anisogamy represents an old and core conflict between the sexes—males' need to provide tiny sperm versus females' need to contribute a large, energy-intensive egg—that resolved in the favor of males.[54] He calls fertilization "genetic parasitism." As Anne Fausto-Sterling argues, there has been a singular focus on the familiar script of sexual difference and male dominance.[55] Patriarchal and cultural modes of sociality got "biologized" into ideas of the "natural" as a scientific fact. Thus, the evolution of anisogamy early on assembled a script of sexual difference and with time continually reinforced ideologies of "compulsory heterosexuality" that we see narrated through theories of floral biology we see today.

- Over 85 percent of angiosperms are hermaphrodites, and many can self-fertilize. They have mixed mating systems with wide variation—from predominantly self-fertilizing to exclusively outcrossing, even varying by population within a species.[56] In biology, sexual reproduction (which produces variation through genetic recombination) is often posited as the explanation for the diversity of angiosperm species and for greater genetic variation that allows for new adaptations to changing environments.[57] But recent work challenges the idea that sexual reproduction always generates recombinant and variable organisms.[58] After all, many asexual species have survived for millennia and still persist.[59] Clonal reproduction is also inventive and adaptive, while sexual reproduction does not always generate variable organisms.

The stories we tell about male and female flowers closely resemble the stories we tell about human males and females—an entertaining, cunning, and crafty game of love, an evolutionary arms race where males try to outwit females and compete with each other. This story, aptly called the "ardent male—coy female hypothesis," is based on a series of experiments on fruit flies by biologist A. J. Bateman.[60] As Marta Wayne summarizes:

> Way back in 1948, Bateman wrote a paper about how often female flies will mate in the laboratory. The paper is used in textbooks . . . as the classic study demonstrating a central tenet of animal behavior, that females will be the choosy or "coy" sex, while males will mate with anything that crosses their path. The idea is that since eggs are larger and fewer than sperm, females make a greater investment in their offspring than males do, so they have to make sure that their fewer offspring get the best possible father. Bateman's data are said to show that a female fly will mate only as often as necessary—once or twice—to ensure fertilized eggs for the rest of her life.[61]

While in the first ten years, the story did not get much traction, it was taken up by some key biologists, especially Robert Trivers, and soon emerged as a bedrock principle of sexual selection, known as Bateman's rule/principle, showing, in Bateman's words, "an undiscriminating eagerness in the males and a discriminating passivity in the females."[62] Bateman's work has come under heavy fire, especially among feminists—in particular Zuleyma Tang-Martinez, Sarah Hrdy, and Ruth Bleier—who together make a thorough, rigorous, sustained, and in my view entirely convincing critique of Bateman's work.[63] Re-examining his work, these feminists challenge his observations as well as his experimental design, methods, and conclusions. They repeated his experiments with different results. Female flies, it turns out, are anything but coy. Yet the imagined sexual landscapes of Darwin's sexual selection and Bateman's rule live on as mainstays of contemporary biology. Darwin proposed sexual selection as a "testing ground of racial character" and a "causal force that could create new races."[64] The logic of Bateman pervades plant reproductive biology, humanizing the bountiful small pollen and the large ovule in the familiar tale of the battle of the sexes.[65]

This sexual script is most evident in the crowning glory of plant sex, the orchid. Orchids embody some of the most fantastic, intricate, and quirky aspects of plant sex. Indeed, Darwin devoted considerable time to his work on orchids, including several papers and books. Evolving over 80 million years ago, orchids display breathtaking innovations—floral structures, fragrances, nectar, putrid smells. The basic story usually told is that orchids and insects have coevolved with each other. Insects are attracted to the flower because of the rewards (such as nectar and pollen) that the flowers offer. In crawling through the flower to find the reward, they are covered with pollen, and when they visit the next plant, they transport the pollen to the new flower, enabling cross-fertilization.

One of the more hotly debated questions comes from the genus *Ophyrs*, where one-third of species offer few or no floral rewards like nectar, and yet are successful.[66] *Ophyrs* is a prolific genus, showing an incredible adaptive radiation with an explosion of new species, highest among angiosperms.[67] For biologists, such high speciation rates beg deeper study. The story of *Ophyrs*, in popular retellings the "bee orchid" (some botanists call it the "prostitute child"), notes that the flowers "resemble" a female bee, practicing "sexual deception" by mimicking the appearance, scent, and tactile experience of a female bee.[68] A male bee spots the flower and excitedly alights, "performing movements which look like an abnormally vigorous and prolonged attempt at copulation."[69] This vain attempt to mate is called "pseudocopulation." In the mating attempt, the male bees (which botanists call "flying penises") get coated with pollen and are thus "sexually swindled." "Parasitizing male behavior" and the "sexual frustration of bees" turn out to be an essential part of the orchid's reproductive strategy. Orchids "strategize" to keep the bees interested. In a given population, each orchid smells and looks ever so slightly different from the others, revealing an extraordinary diversity of orchid "mimicry."[70]

It is striking to see botany's inability to capture this extraordinary tale except as a tale of deceit and falsity for the bee. Ironically, the story is told as if the orchid is a deceptive female, when the plant has both pistils and stamens. While Victorian sexuality punctuated much of the botanical imagination, colonial discovery of orchids transformed them into exotica in western gardens and living rooms. Orchids are forever saturated with racialized and oriental scripts of the erotic, beautiful, strange, dark, alien, dangerous, and mysterious. The sexual narratives are dramas filled with sensuality, subterfuge, and sexual excess. While the species' rich genus has without a doubt evolved intricate, sensual relationships with their bee mates over millions of years, the botanical imagination reduces it to the narrow vision of Eurocentric sexuality, albeit sometimes in a version of the exotic Orient.[71] In narrations of this story, the limits of anthropocentrism are apparent, including how unsexy our language of sex really is. Surely we can retell the amazing diversity of flowering plants in less human-centered language without collapsing it into a Victorian morality play of lascivious gentlemen and coy ladies, a tale of the battles and wars or oriental allure?

At the heart of the problem is a more profound irony. In the orchid's wet dream, the orchid does not dream of the opposite sex; it dreams of the bee's sensual presence in its lush folds, luxuriating in the sweetness of its nectar, the stickiness of pollen, and the wild, glorious sex with its erotic visitor.

Act III: Playful Intimacies: Do Plants Have Sex?

How do we move beyond Eurocentric tales of eager gentlemen and passive ladies or the canny, seductive flower and the hapless male who cannot resist its beauty? Let us look more carefully at the reproductive lives of the orchid we just described.

Some questions we should ponder in the naturalized narrative that biologists tell: When plants self-fertilize, is it incest? When bees copulate with plants, are they having sex? Are they masturbating? Why do we not refer to human masturbation as pseudocopulation? Do we even have a vocabulary for discussing and thinking about nonreproductive sex except as failure, perversion, defective, and useless? Despite continued debate, why do we extoll sexual reproduction as the pinnacle of evolutionary adaptation, and indeed the only reason for sex?

Do plants even have sex? Males and females in plants never meet, touch, feel, or engage. What is striking in our anthropomorphic tales is when we choose to anthropomorphize and when we choose to leave inconvenient details aside. If sex is about union, closeness, attraction, intimacy, contact, engagement, feeling, and pleasure, who exactly is having sex?[72] Flowers are intimate with insects such as bees, flies, wasps, beetles, butterflies, ants, and with birds and bats. These intimacies have been fostered through millions of years of coevolution. They are also intimate with the wind as pollen glides in its currents and with water as pollen floats in its streams. The ovule has no contact with the flower from which the pollen originates; there is always one degree of separation. Even when plants self-fertilize, there is no intimate contact.

If the flower is intimate with the bee, wasp, fly, bird, wind, water, and earth, should we think of the doctor, the nurse, the sperm bank, the egg bank, and the turkey baster—the vast infrastructure of assisted reproductive technologies—as sexual partners? Why ever not? Or perhaps the bee, wasp, fly, bird, wind, and water represent the assisted reproductive technologies of flowers? If we are being anthropomorphic, perhaps we can be thorough about it. Part of the issue is that we anthropomorphize only when the details validate and confirm certain stories. These elisions and erasures are central to our stories, our storytelling modes, and the sexual scripts they elicit. In such an opening up, we might retell the bountiful, glorious, promiscuous, and perverse sex of the natural world– of flowers and humans alike.

An ableist vocabulary in biology laments the sessile, grounded, immobile plant. Yet this cursory overview of plant reproduction shows a wildly agentic and mobile plant world. In fact, pollen and seeds of plants travel far and wide. Perhaps not in the ambulatory mode of humans or animals, but why must our ideas of mobility be based on the ableism of human worlds? Arguably, the evolutionary strategy of plants has been one that far exceeds and likely will outlive humans. Ultimately, it is not about movement, or mobility, or ability in a human sense. It is about a different vocabulary, one we have yet to develop to adequately capture the world of plants. We need vocabularies that capture their playful, affective worlds that build community and intimacy with the world around them. We need to capture the global reach of pollen and seeds as they travel with the flows of the winds, the waves of water, the beaks of birds, the legs of bees, the fur of animals, and the pull of gravity and are assisted by the power of fire and the wetness of earth. We need to better account for plants and their naturecultural entanglements with the human—their exploitation through the circuits of colonialism and through being traded in a globalized world.

Evolutionary biology and Darwinism have remained grounded in cis-heterosexual reproduction as the critical fulcrum of natural selection. But is heterosex everything? Is there room for other possibilities? Can we narrate other stories beyond the suffocating tales of a rationalizing and naturalizing patriarchal culture grounded in male sexual violence? At the heart of Bateman and Darwin is the primacy of sexual reproduction that drives the heart of evolution. What matters in biology is the "fitness" of the individual, which can be measured only post hoc by the number of their successful progeny, not by any other quality of the organism.

In highlighting the reductive logics of cis-heterosexuality, we see the limited languages of not only sexuality but also community, intimacy, and belonging. The point is not whether plants have sex as we understand it. Rather, the point is our continual obsession with sex and our inability to think and talk about the biological without the language of sex. How might we learn to read, in our scientific interactions with plants, the many delights and wonderous biologies of plants tales beyond sex and sexuality, tales of other intimacies, other belongings, and other ways of living?

Conclusion: Enabling Floral Dreams

Colonization is not only an act where land is stolen; also stolen are words, language, culture, histories, theories, knowledge, forgotten sexual imaginaries, landscapes of affective ecologies, and other modes of sameness and difference, or what McKittrick evocatively calls *livingness*.[73] My biological education destroyed the magical *livingness* of childhood. India, like others, may have emerged a postcolonial nation, but its biological education resituated a colonial script.[74] As Leanne Simpson powerfully notes, the plunder, pillage, and desecration, the ravaging of colonized lands and people, was followed not only by erasure and forgetting but by a rationalizing and retelling of colonialism as a civilizing mission—in the histories by the colonists and in the psyches of the colonized.[75] Colonial botany and its retelling of the workings of the natural world (separated from the cultural world) ground the contemporary biological tales of plant reproduction. While colonialism is central to the emergence and development of the science of botany, this retelling allows a narration of botany as removed from and unconnected to its colonial roots. The breathtaking, heterogeneous, and multitudinous cosmologies of indigenous people across the world were replaced by a universal science in a foreign tongue—a cosmological scheme forced not only on humans but on all planetary life.[76] A plethora of languages, theories, terminologies, cosmologies, spiritualities, and sexualities were reduced to a theater of Victorian gentlemen and ladies. What a loss!

Colonial imaginations pit the universality of western botany against the rich contextual specificity and contingency of a multitude of local and indigenous knowledges. Yet even within western botany, there are other imaginations. For example, Joan Roughgarden not only rejects Bateman's rule but challenges

Darwinian sexual selection theory altogether, replacing it with "social selection,"[77] recognizing the vast number of "exceptions," of behaviors that do not conform to Darwin's and Bateman's cis-heterosexual imaginary: monomorphic species, sex role reversals, homosexual mating, trans and queer animals, and so on. Recent work offers alternatives to the obsessive focus on sex as copulation and intercourse, opening up life as vast infrastructures and landscapes of sociality—not only the language of competition but languages of affinity, cooperation, mutualisms, and symbiosis.[78] The world is teeming with multispecies engagements that require more narrative worlds than the reductive mode of mainstream biological storytelling.

Lynn Margulis and Dorian Sagan go one step further in arguing that sex is a feature inherited from ancestors, an "imperative relic" inherited through eukaryotic ancestry.[79] They link sex to the accidental evolution of multicellularity and cellular differentiation, of which sexual reproduction was a byproduct, which subsequently became inextricably linked. As Fausto Serling argues, for many scientists the adaptationist "just so" story emerges because it is assumed that "sex must be good for something or we would not have it."[80]

Queer ecologists remind us that "sexual and gender becomings [are] complex biological, technological, and political assemblages rather than . . . either purely discursive or biologically determined processes."[81] The wide landscape of sexuality, including asexuality, homo/heterosexuality, and pansexuality, challenges us to not only embrace multiplicity but also question the idea of sex as always obviously pleasurable.[82] Evolutionary biology is aligned with queer theory if we follow the messy terrain of adaptations that produce astonishing variation in plant reproduction. How we narrate plant sex reveals how deeply entrenched these sexual scripts have become—told through the biologies of all kinds of plants and animals. The problem is a willful refusal to appreciate the asexuality of some beings and the sociality of sexy interspecies entanglements as a way to radically open up questions of what sex is and can be. Rather than always desexualize plants or eschew sexual metaphors, we can embrace the promiscuous, playful perversity of plants to multiply, obfuscate, and challenge the blurred boundaries between human and nonhuman, animal and vegetal, sexual and asexual. We could also challenge landscapes of pleasure and sex altogether as prerequisites for a biological life. Plants offer many queer possibilities. We can embrace them all!

In anchoring Toni Morrison's ethics of rememorying in vocabularies and prisms of plant reproductive biology, we see the vivid imprints of colonial violence. Reproduction ensures an endless cycle of (invisible) gendered and colonized labor (of humans and plants). Rememory insists that we take responsibility for our botanical amnesia. It opens up the many cultural imaginations and cosmologies destroyed, ignored, marginalized, or rendered invisible by colonialism.

We do not have to reject the particularisms of cultural and local understandings for the universalism of science. One of the key insights of recent work in feminist STS has been to challenge binaries. The binaries of nature/nurture, biological determinism/social construction, genes/environment, nature/culture, and sciences/humanities are no longer adequate explanations.[83] It is not that we need to take into account both sides of the binary to arrive at some happy medium; it is that each category is dependent on the other. Nature and culture are not binary opposites but rather co-constituted. We need new interdisciplinary methods that allow us to understand how scientific knowledge has been produced, and even more methods to make a naturecultural world and its workings anew. For this, we have many genealogies and landscapes to work with—Sylvia Wynter's counterhumanism, Gloria Anzaldúa's borderlands, Audre Lorde's erotic, Angie Willey's biopossibilities, Katherine McKittrick's livingness; Rebecca Herzig's fundamental unruliness; Octavia Butler's radical fiction; Donna Haraway's speculative fabulations; Chandra Mohanty's feminism without borders; and Muñoz's queerness as limitless possibilities.[84] Liz Grosz, Myra Hird, Carla Hustak, Natasha Myers, and Elizabeth Wilson remind us that queer imaginations also reside in Charles Darwin's orchids, barnacles, and teeming, entangled banks.[85]

Sexuality as we know it—the normative world of cis-heterosexuality—is not the body or an act but an ideology of western colonialism. In asking why and how plants have sex, we are left asking whether they do or must. Only in thinking through the long arc of colonialism and understanding the histories of Linnaeus and beyond have I have come to appreciate the profound androcentrism that grounds scientific views of plant reproduction. Thinking back on my biological education, I don't think plants have "sex." It takes many elisions and erasures in our stories about sex, sexuality, race, and reproduction, many analogies and sleights of hand, to make plants into European gentlemen and ladies.[86]

CHAPTER SIX

In the Dark Shadows of the Tree of Life

Sexuality, Race, and Reproduction

I am not nostalgic [for a country which doesn't yet exist on a map]. . . . Belonging does not interest me. I had once thought that it did. Until I examined the underpinnings. One is misled when one looks at the sails and majesty of tall ships instead of their cargo.

DIONNE BRAND | *A Map to the Door of No Return*

The female body in the West is not a unitary sign. Rather, like a coin, it has an obverse and a reverse: on the one side, it is white; on the other, not-white or, prototypically, black. The two bodies cannot be separated, nor can one body be understood in isolation from the other in the West's metaphoric construction of "woman."

LORRAINE O'GRADY | *Olympia's Maid*

I WAS TAUGHT TO BE GRATEFUL FOR TREES. In my botany classes, I discovered their indispensability to life on earth, their photosynthetic living releasing life-sustaining oxygen into the atmosphere. In my everyday life, I enjoyed how the aromatic scents and vibrant colors of their flowers gratified the senses. Their leaves, flowers, and fruits fed and nourished me. On many a hot summer day, their shade offered comfort.

Wondrous Indian mythological stories of my childhood were filled with magical trees. Today, in an era of climate change, I am indebted to their ability to sequester carbon. I had not contemplated the dark side of trees until I heard Billie Holiday sing Abel Meeropool's "Strange Fruit." The sensuous poetry, the shocking imagery, the haunting melody, and the ethereal voice together discombobulated me, literally and viscerally. Listening to Holiday, I glimpsed the horrors of lynching; the song evoked the violence not only of the body but also of the spirit.[1] These days, I hear the haunting spirits of "Strange Fruit" as I read about the rise of lynchings in India, the growing violence and intimidatory tactics against minorities, particularly Muslims and Dalits. With time I have come to understand that violence is more than the horrific reminder of the long, brutal histories of slavery and colonization. In their afterlives, violence is rehearsed, again and again. They live on; they endure. As Sylvia Wynter reminds us, this history bespeaks the critical logics of empire that form the infrastructure of our political formations. They ground and shape western knowledge systems and undergird contemporary academic disciplines. "Race" in Wynter's conception is central to western humanism, a foundational anti-Blackness we can trace to 1492. I seek to understand how logics of empire, translated into theories in the study of biology, in turn shaped colonial and biological infrastructures of racialized sex. I think through the history of plants and colonialism in the context of feminist work on reproduction.

Plants, like all objects, are caught up in naturecultural histories. Flowers like pansies, green carnations, lavenders, and sapphic violets are decidedly queer.[2] In the dazzling array of colors in daylilies, plants with darker-colored flowers were given names referencing Blackness, while light and white colors were not racialized.[3] The cultivation of flowers is also caught up in the transnational politics of racial capitalism.[4] Here, I want to focus less on the embrangled naturecultural histories of plants and more on the deep history of race, sexuality, and reproduction. I want to think in the dark shadows of trees. In this light, so much of the history of botany is cast anew, yet hauntingly familiar. In botany, sex and reproduction are foundational, the theoretical and biological core to explain the splendor of life on earth. Thinking in the dark shadows of trees, I now see how sex, sexuality, and reproduction are also fundamentally racialized as variables, structures, and indeed infrastructures of domination.[5]

The Coloniality of Gender

To understand the conceptual power of normative sex and sexuality, I turn to the work of feminists who have highlighted how race, sex, and gender underlie coloniality, including Maria Lugones, Sylvia Wynter, Breny Mendoza, Catherine Walsh, Freya Schiwy, and Kiran Asher.[6]

Echoing and extending Quijano's argument, Lugones argues that as a Eurocentric, global, modern colonial gender system cohered and global capitalism was constituted through colonization, "gender differentials were introduced where there were none . . . the imposition of this gender system was as constitutive of the coloniality of power as the coloniality of power was constitutive of it."[7] Lugones argues that the colonization of the Americas and the Caribbean imposed a hierarchical distinction, including gender differences between men and women. Gloria Anzaldúa declares, "I know things older than Freud, older than gender."[8] Scholars in indigenous, decolonial, Black, and trans studies critique enlightenment theories of sex and gender, largely produced by white Anglophone scholars.[9] They recontextualize our understandings of gender and sexual difference within the racial and colonial system of the violence of slavery and conquest.[10] In the colonial narrative, only "civilized" people (Europeans) are human. In contrast, the colonized were deemed animalistic and thus "non-gendered, promiscuous, grotesquely sexual, and sinful."[11] Lugones highlights how race and gender are deeply entangled concepts; hierarchies of colonizer/colonized blur into the human/nonhuman and the human/animal.[12] Colonized subjects became the mythical Others, dehumanized and animalized. Most scholars today agree that while diverse landscapes of sex and gender emerged through colonialism, a normative European conception of a sex/gender/sexuality system was weaponized into the rest of the world and biology.[13]

This discussion is important to the history of botany because plant reproductive biology is imagined in human terms, including terms such as *sex*, *gender*, and *sexuality*. As Greta LaFleur argues, when Linnaeus added humans to his taxonomy of Mammalia, sexual characteristics putatively endemic to each race were used to illustrate racial difference.[14]

Before we move to plants, let me summarize the work on race and colonialism. Although many scholars have explored this topic, some key issues are worth highlighting.[15] Arguing that the colonial imposition of gender goes far beyond our understandings of patriarchy and heterosexuality and the control of sex, Lugones makes an important distinction between the light and dark side of colonial gendered frameworks. The "light" side of the gendered framework characterizes the normative European "human" into two dimorphic sexes organized through a complementary heterosexuality. In contrast, the "dark side" portrayed the colonized as hypersexual and hermaphroditic. While colonizers were sexually dimorphic, the colonized were ambiguous in their sexuality: "sexual fears of colonizers led them to image the indigenous people of the Americas as hermaphrodites or intersexed, with large penises and breasts with flowing milk."[16] Linnaeus characterizes African women (*Africanus Niger*) as women "without shame," a representation that becomes synonymous with black femininity as imagined by humoral (body fluids) associations with Africa and its geography and climate.[17] Global histories such as the case of Sara Baartman, a Khoikhoi woman who was exhibited as a show attraction in nineteenth-century Europe, reveal complex colonial social relations and representations of anatomy in the "Hottentot Venus."[18] In her canonical essay, Evelynn Hammonds likens black women's sexuality to black holes, seemingly empty but in fact dense and full.[19] The body (especially genitalia) emerges as central in this racial understanding: "Sex becomes an origin story. This use of sex allows for naturalizing claims of inferiority and superiority across colonial lines."[20] To complicate matters, gender and sex are elided. As Mel Chen argues, "The 'genitals' are directly tied to social orders that are vastly more complex than systems of gender alone."[21]

Grounding difference—race, sex, gender—in the body created stark differences between the (often animalized) enslaved and the colonizers.[22] These distortions of historical western intellectual thought, especially the biological sciences, track into plant and animal worlds. Alexis Pauline Gumbs's majestic *Undrowned* weaves the submerged wisdom of marine life with histories of slavery and incarceration.[23]

The concept of *coloniality* thus helps bring together the many regimes of the colonial. It allows us to understand the decimation of indigenous societies across the world, the birth of settler colonialism, and the postcolonial condition.[24] Histories of racial capitalism are linked by the extraction of human and plant resources and labor. Mamdani recounts how while the Spanish Empire encroached on the Americas, they were also involved in the ethnic cleansing of Moors and Jews. In a postcolonial understanding of modernity through the Americas and 1492, we see that nationalism and global colonialism are co-constituted.[25] Coloniality and its civilizing mission are a global feature of European colonialism—not only in the Americas but also in Africa and Asia. It helps us understand how colonial legacies shaped and continue to shape the field of science, particularly botany.

The Tree of Life

The figure of the tree looms large in biological classification. This is not the biblical tree of life but Darwin's famous tree of life, which introduced his theory of descent with modification.[26] His simple branching tree has blossomed into complex mathematical and technological theories.[27] Exactly what form the tree takes—how many trunks and branches it has, the pattern of branching, and so on—depends on the classificatory model of evolution and its methods and philosophies. Whatever the model, evolutionary trees are branching structures, where the ancestral individuals lie in the nodes and the latest evolutionary divergences at the tips.

Evolutionary theories have had a tense relationship with theories of nomenclature and classification. Linnaeus looms large in any discussion of nomenclature and classification. His binominal classification (genus and species) expanded into a trinomial (genus, species, and subspecies). In refining classification beyond the genus and species level, the subspecies or "race" emerged. Linnaeus used the term *varieties*, and many thus view Linnaean varieties as subspecies or "races."[28] For Linnaeus, human races emerged from divergent geographies that endowed groups with profound differences in physical, moral, and intellectual capabilities.[29] Today, we think about race largely as a category of the human, but race in fact emerged as a category for all organisms.[30] In the 1950s, many scholars critiqued and questioned the idea of the subspecies (or race) as a useful biological category.[31] Even before then, many, including Darwin, believed that subspecies and categories like varieties, geographic races, and demes were too ephemeral and the category too artificial and arbitrary to be useful in biology.[32] Race has since largely disappeared as a category within plant and animal classification, but thanks to enduring colonial politics, it remains significant in its avatar in the human, only to travel through biological concepts back to plant worlds.[33]

I focus on the tree of life because it is metaphorically and theoretically important. Metaphorically, the tree of life allows claims of similarity and difference to coexist since we all inhabit the same tree, while some of us are superior to others. Histories of monogenism and polygenism, and their attendant racialized histories, are all part of theories of evolution. I begin with foundational theories from the feminist studies of race, sexuality, and reproduction, then think through the histories of biology and botany in particular. I end with some reflections on the colonial histories of sex and race.

Feminist Studies of Race, Sexuality, and Reproduction

Sexuality is foundational to academic disciplines.[34] Terms such as *race*, *sex*, and *sexuality* have been deeply interconnected in debates and ideologies since the eighteenth century.[35] Histories of slavery, indentured servitude, and settler colonialism powerfully control the "politics" of reproduction. The legal doctrine *partus sequitur ventrem*, or "that which is born follows the womb," ensured that children of slave mothers would inherit the legal status of the mother.[36] Colonial logics and scripts created infrastructures of control—*coloniality*—shaping biological theories of sexuality and reproduction, subsequently shaping the biology of plants and humans.

SEXUAL BIOLOGY

Scientific theories of biological reproduction, especially theories of sexual reproduction, emerged and solidified during European colonial expansion. But these were racialized. Here, two sets of representations are important. First, Linnaeus remade plants into humans, not only in attributing them with a binary sex but in endowing them with qualities of European husbands and wives.[37] But like colonized and othered humans, plants did not always follow the monogamous visions of empire.[38] Second, the laboring body and its reproduction, plant or human, were critical to colonial needs. Controlling and molding sexuality played an important role in supporting colonial ambitions.

The idea that categories of sex and sexuality are coconstituted through the category of race and through the histories of colonialism is well explored in human history, but less so in botany. Logics of biological inferiority are critical to colonial exploitation. Under colonialism, animals, plants, and the land were rendered insentient commodities. They were there for the taking and making, for exploitation and extraction. The tree of life connected all life on earth and portrayed some humans as inferior. Colonial discourses of inferiority provided the scientific rationale to commodify colonial humans—they were not seen as "human," and if they were human, then they were a subspecies or race lower than their European counterparts. This meant that they could be commodified, bought, sold, and experimented upon.

Hierarchies of sex differences are always racialized. As I argue in *Ghost Stories for Darwin*, multiple lines of evidence urge us to understand sex as a racialized variable. Nancy Stepan demonstrates how repeated and direct analogies were made between identity categories.[39] For example, the race-gender (read race-sex) analogy was used in the nineteenth century to claim that women's smaller brains and protruding jaws were evidence of their evolutionary inferiority to men. By analogy, a vast swath of humanity seen as inferior—women, so-called lower races, the sexually deviate, the criminal, the urban poor, and the insane—were constructed as a biological "race apart."[40] By the mid-nineteenth century, racial biology became a science of policed boundaries. In a similar argument, in "Pelvic Politics," Sally Markowitz argues that the very idea of sex differences emerged not from binary differences of gendered bodies but from racialized meanings of sexual bodies. Markowitz traces the category and ideology of sex/gender to show how it was based not on a simple binary opposition between male and female but rather on a scale of "racially coded degrees of sex/gender difference."[41] Evolutionary biology produced the "manly European man" and the "feminine European woman" as evolutionary types that marked the pinnacle of a hierarchical project of race.[42] Thus, characteristics like femininity are themselves racialized in the historical record; "femininity" could not be assumed to be a characteristic of *all* females.

What emerges from such different ontologies of race and sex is a "logic of difference" that underlies the foundations of medicine and science in creating racialized and gendered bodies.[43] For example, in tracing the origins of gynecology, historians have shown how racialized understandings of black women as "strong" (as opposed to elite white women's fragility) translated into an extraordinary biological capacity to bear pain. Thus, the "father" of gynecology, Marion Sims, performed surgeries on black women without anesthesia. Their sameness ensured that the knowledge gained was transferable to white women; their difference ensured torture. The logic of difference always benefits the elite and powerful at the expense of the poor and marginalized.[44]

Controlling Biology, Controlling Reproduction

While colonized people were relegated to being inferior humans (or nonhuman animals), they were certainly not plants. Each organism had its place in the great chain of being. My exploration of the logics of science is not organized around finding any easy analogy between the colonized peoples and plants. Rather, the logics and needs of empire were translated into universal theories in biology that affected plans and humans. Here, I think through plant colonialism in the context of feminist work on reproduction.

REPRODUCING THE BODY

Reproduction is central to the politics of colonialism, slavery, and indentured servitude—promoting the reproduction of "objects" with value while controlling (or eliminating) the reproduction of others. If plant classification brought order to the natural world, colonial biology brought control into it. Science is critical to this story. Within human history, control is exemplified in histories of eugenics, scientific racism, and sterilization abuse. Eugenic scripts determine who reproduces, who lives or dies. Saidiya Hartman eloquently summarizes slavery's dominating power over life:

> I was determined to name and articulate the character of this power, which was an assemblage of extreme domination, disciplinary power, biopower, and the sovereign right to make die. The dimensions of subjection traversed the categories of human, animal, and plant. Slavery's modes of accumulation and exploitation failed to be explained by precapitalist modes of production or the factory floor. The character of gendered and sexual difference, and negated maternity and severed kinship, bore no resemblance to the intimate arrangements of the white bourgeois family and cast out the enslaved from the nomenclature of the human.[45]

As Laura Briggs persuasively argues, all politics is reproductive politics.[46] Historians of reproduction and race have, by now, assembled a powerful account of how extractive ambitions of colonialism and slavery shaped the politics of reproduction—the coloniality of reproductive logics.[47] Whether one considers questions of demography, labor, medicine, science, work climate, welfare, food and nutrition, or equity and equality, reproductive politics loom large.[48]

What can these key ideas from the history of sexuality and reproduction teach us? Here I explore the history of botany through the feminist literature. Five critical themes shape this comparison: the commodification of bodies and reproduction, the extractive logics of empire, the determination of desirable and undesirable bodies, how science imagines the world as a Eurocentric petri dish, and finally the politics of deanimating life.

PLANT COLONIALISM: CONTROLLING PLANTS

Not all plants are equal in the colonial imaginary—they must earn their place by capturing the colonial gaze: aesthetically pleasing, tasty, useful, medicinal, economically valuable, cultivable, profitable, or exotic. At the heart of plant colonialism is plant labor. Agricultural infrastructures necessitate that all forms of human and nonhuman viability be driven by maximized yield.[49] As long as plants know their rightful place as laborers, providers, and controlled commodities, their positions manipulated and controlled, their presence is tolerated. Once they are accused of unruly practices that prevent them from staying in their subservient place, they threaten the natural order of things.[50] Industrial plant cultivation is a project of domestication. Yet agriculture, the culture and cultivation of plants, has many possibilities, the colonial model being very different from other scientific and indigenous models across the world.

A central objective of colonialism was the commodification of bodies and reproduction. To turn objects into property that can be owned, manipulated, bought, and sold necessitates a vast infrastructure of laws, trade routes, governance, labor practices, and brutal regimes of control. These logics shaped the histories of slavery, indentured servitude, and colonial rule. As Weinbaum argues, human commodification must be "understood as subtended by the long history of slave breeding as it was practiced in the Americas and the Caribbean."[51] Historians of slavery and colonialism have painted a vivid portrait of the inhumanity of colonized life.[52] Since emancipation, the afterlives of slavery continue to reside in the intergenerational trauma of the living, in the bones, flesh, and blood. We have seen this repeatedly during health crises, as class and race shape patterns of mortality.[53]

A similar biopolitics of gender and race plays out in plant worlds. Anthropologists remind us that humans once foraged for food. They traveled to find plants and then harvested what was needed but secured the original source for the future. In many cultures, people give thanks to plants and worship their life-giving properties.[54] Indeed, much of biological evolution is a story not of lone species adaptations but of coevolutionary forces. This is no simple tale of exploitation. Plants have evolved to count on humans, insects, birds, and other creatures to eat their fruit and help with seed and pollen dispersal. In the biological literature these have been characterized in terms of cooperation, mutualisms, symbiosis, or mutual transactions.

What fundamentally changes with the advent of colonialism is the scale of production, the creation of universal regimes of control, and often short-term profit logics. The domestication of plants and the sciences of plant breeding, hybrid crops, and transgenic plants have changed the story of mutual exchange into one of increasing control over the biology of plants. Industrialized agriculture is entirely about control and commodification, increasingly enhanced through technology. Plants are chosen for their productive and reproductive value—the ideal plants are fast-growing, transportable, and long-lasting products. Often, once they were found, they were bred into monocultures. Flexible genetic bodies, botanies of "desire," lent themselves to extraordinary replication—corn, rice, wheat—forming the bedrock of commercial crops.[55] Consider this: With over 170 potential crops, only three staples provide 40 percent of the world's food energy intake.[56] As Robin Wall Kimmerer astutely sums up: "In indigenous agriculture, the practice is to modify the plants to fit the land. As a result, there are many varieties of corn domesticated by our ancestors, all adapted to grow in many different places. Modern agriculture, with its big engines and fossil fuels, took the opposite approach: modify the land to fit the plants, which are frighteningly similar clones."[57] Colonial logics have ushered in technologies of increasing control. My argument is about colonial and capitalist logics, not technology. Many indigenous cultures and plant scientists work with plant worlds without such logics. After all, plant breeding is a very ancient technology.[58]

The short-term profit motive is key here. The extractive logics of empire, hallmarks of colonial forestry and global plantations, endure in contemporary industrialized agriculture. For example, while farmers once saved seeds for the next year, they are now sold hybrid and transgenic seeds that force them into dependence on seed companies. Worries about the dominance of megacorporations that promote high-input agriculture are warranted. For example, in the 1990s Monsanto developed "terminator" technology. The rather ingenious but perverse technology programmed cell death into the seed. So while the farmer could grow the crop for the year, the seeds were not viable, forcing farmers to buy seeds each year.[59] Agriculture, the science of life, is now marked by technologies of death.[60] Such practices have created a kind of modern seed sharecropping with its own debt-ridden futures. Ultimately, after much uproar, Monsanto "temporarily" set aside the technology.[61] While some countries have banned such technologies,[62] such histories engender deep suspicion over technology rather than companies. High-input agriculture and its ecology of seeds have devasted ecologies the world over. But as Donna Haraway reminds us, technological innovation is important, and we should be open to it even while we resist and refuse proprietary technologies and secretive regimes of patent exclusivity.[63]

In a feminist frame, a history of colonial plant breeding is a history of violence. With the rise of botany, all agency and claims of sentience were erased from plant worlds. Through a colonial gaze, plants existed for humans—to be moved across borders, culled, exterminated, in/outbred, modified, endlessly manipulated. The predominant colonial script is about choosing and cultivating varieties as efficiently as possible. So much of this logic explains our culinary habits today, especially the increasing homogeneity of the global food system.

Racial capitalism has been one of the more useful frameworks for understanding the histories of colonialism and slavery. At its core, racial capitalism refuses the idea that capitalism can ever be understood outside of racial formations.[64] The extractive logics of empire, of humans and plants, were fundamentally racial. As Lisa Lowe argues, racial capitalism's expansion through extractive commodities and economies were never uniform but always strategic by seizing upon colonial divisions, selecting certain regions for production and others for neglect, and using certain populations for exploitation and others for disposal.[65]

Plants are also trapped in the logic of racial capitalism. As Xan Chacko powerfully argues, through colonial technologies such as Wardian cases (small wood and glass boxes that helped transport live plants) or seeds in envelopes that were sent via mail, "the reproductive bodies of plants have been extracted, commodified, reproduced, and proliferated to satisfy human needs and desires."[66] These goals were achieved through colonial hubris and reductionist thinking in biology. During colonial rule, colonists transported plants wherever they went, creating "little Europes" along their routes. Shared information and logics shaped plantations around the globe. Even in places with successful independence movements, the western-centeredness of the biological sciences in the postcolonies remains firmly in place in agricultural, research, and educational practices. With respect to the plant sciences, the west remains the site of scientific innovation.[67] Innovations produced in laboratories in the west became exemplars for solving the world's problems. Scientific research, increasingly the handmaiden of corporations, is no longer "free." As we saw earlier with genetically engineered seeds, protected by patents owned by western or multinational companies, third-world agriculture is caught in endless debt loops. Contemporary farmer social movements work to challenge colonial logics, even in the postcolonies.[68]

Technology, often posited as a solution, is also caught up in colonial logics. Natali Valdez makes this point well and powerfully.[69] When the science of epigenetics emerged, feminists saw promise in a renewed focus on the environment, where responsibility for fetal and infant health would not be reduced to the pregnant body. But as Valdez shows, old eugenic scripts were repurposed to blame powerless women (especially poor women and women of color) for historical and ongoing social traumas. We see the same in agriculture. We produce drought- and flood-resistant crops rather than attend to climate change. In short, the history of capitalist medicine and industrialized agriculture satiates consumer desire through technological fixes rather than rethinking logics of control.[70] Agricultural innovations such as monoculture crops, plant breeding, genetically modified seeds, herbicides, and pesticides have decimated local ecologies through overuse of water resources and soil nutrients. Environmental concerns draw elite consumers not to dismantle the inhumanity of agriculture but toward movements for organic seeds and low-input agriculture, spawning its own logics of purity of savior seeds, seed banks, and regimes of care. We are stuck in an endless loop of technological determinism. Each technological failure is met with a technological fix.

The long arc of ecological colonialism has transformed planetary landscapes into lawns, monocultures, plantations, and commercial forestry. This exploitative history is countered by a pure "nature," relegated to conservation lands, public parks, national parks, and remediation sites. It is a politics of purity through control, with endless cycles of destruction and remediation.

Extractive logics have spawned ecological destruction. Take, for example, the environmental impact of the much-heralded green revolution (also called the Third Agricultural Revolution, from after World War II through the 1980s) that celebrated food self-sufficiency in India. Decades later, the consequences are sobering.[71] We see rising soil degradation, diminishing water resources, and increasing chemical inputs.[72] The green revolution in the state of Punjab, heralded as the breadbasket of India, saw a shift from the use of diversified climate-adapted crops into one predominantly focused on rice-wheat rotation with much higher water use. This has led to a severely depleted water table, and the once fertile state now faces desertification.[73]

Finally, a critical project of colonialism was deanimating the world. Queer and indigenous scholars have argued that in rendering people and other organisms into commodities, exploitation proliferated.[74] As the natural world was deanimated and degendered, plants became nonsentient, nonsacred creatures, free to be felled, moved around, and monocultured, exploitable, killable commodities devoid of rights.[75] Much work in indigenous studies, as well as more recent scholarship in critical animal studies, critical plant studies, and queer ecologies, seeks to reanimate the natural world.[76] Here, like many others, I am less invested in "regendering" plants than in "reanimating" them to foster a more capacious vegetal (and human) imagination.

PLANT STUDIES AND THE NEW TREE OF LIFE

In recent times, especially in the United States and Europe, there has been a proliferation of academic and popular works on the biology of plants. Refusing any mark of human exceptionalism, these works recast plants from the once nonsentient, nonhuman organism into organisms endowed with human qualities. They are intelligent beings with feelings and familial relations, mothering their communities, creating alliances with other organisms, and understanding life and death. They have senses—feeling, reacting, communicating, engaging, and thriving. They help one another and warn others of danger. I am not interested in the science behind such claims or whether these attributes are "true," but I wonder why such representations have suddenly proliferated. These works carry titles such as *Venerable Trees*, *Hidden Life of Trees*, *Wise Trees*, *The Overstory*, *The Ancient Magick of Trees*, *In the Company of Trees*, *The Living Wisdom of Trees*, *Finding the Mother Tree*, *To Speak for the Trees*, and *How I Became a Tree*. I am broadly calling this genre of literature "tree love."

Why is so much "tree love" arising now? Some see it as an antidote to the exploitative colonial histories we just traced. In reflecting on why millions of readers and viewers have become "magnetized by the hitherto arcane field of plant communication," Rob Nixon argues that it results from a quest for "alternate modes of being to neoliberalism, modes more accommodating of the coexistence of cooperation and competition in human and more-than-human communities."[77]

Thinking about human and plants together, is the rise of tree love related to the rise of Black Lives Matter (BLM), Standing Rock protests, and indigenous and trans movements alongside the rising urgency of climate change and other ecological crises? Yet at the same time, many parts of the world have also seen the rise of right-wing nationalist (including white nationalist) movements. Here, Lauren Berlant's insightful theory of the genre of "crisis" as an impasse, not an event, is useful:

> Often when scholars and activists apprehend the phenomenon of slow death in long-term conditions of privation they choose to misrepresent the duration and scale of the situation by calling a *crisis* that which is a fact of life and has been a defining fact of life for a given population that lives it as a fact in ordinary time.[78]

What the explosion of attention to climate change and the "tree love" literature ignores is work from postcolonial and indigenous ecologists who have tracked the long arc of the climate crisis emerging over centuries. For many in the world, the crisis is not new but an ongoing and enduring one. Similarly, the idea of trees as sentient beings, while new to "crisis" worlds, is very much a part of ecological thinking in many cultures of the world. What is striking to me about "tree love" is the whiteness of the genre, which relies on a kind of purity politics. In these west-centered narratives, trees live vibrant "human like" life outside human worlds. There is drama and intrigue, happiness and sorrow. In many cases, the solution is to keep humans away, to protect nature as separate and pure. To be protected, nature must escape humanity. Not surprisingly, absent from these narratives and settings are the long histories of race and colonialism that ushered in the engines of climate change.

The genre of "tree love" emerges from the long history of the botanical sublime, a separation of nature and culture. Here plants are remade but in a human image. I would suggest that in contrast, indigenous, postcolonial, and queer ecologies are thoroughly embrangled, never pure, and constantly making and remaking one another.

Unnatural Selections

If the histories of western science and its natural selections have ushered in landscapes of ecological devastation, how might we take on the challenge of imagining worlds anew? While we have attended to the racialized nature of sex, sexuality, and reproduction, we've focused less on the ways they center the cisgender experience and heterosexuality. Here feminist and queer ecologies inspire. Dorothy Roberts reminds us that race was invented as an instrument to promote racism, not the other way around.[79] In a similar vein, as historian of science Beans Velocci reminds us, sex is "fundamentally incoherent": "It's not transness that's made up. It's cisness that takes a ton of work to create and maintain."[80] In recognizing sex and race as incoherent categories, how can we reimagine and retheorize plants in more capacious and imaginative terms? In discussing animal sex, Sharon Kinsman's thoughts are useful:

> Because most of us are not familiar with the species, and with the diverse patterns of DNA mixing and reproduction they embody, our struggles to understand humans (and especially human dilemmas about "sex," "gender" and "sexual orientation") are impoverished . . . Shouldn't a fish whose

gonads can be first male, then female, help us to determine what constitutes "male" and "female"? Should an aphid fundatrix ("stem mother") inform our ideas about "mother"? There on the rose bush, she neatly copies herself, depositing minuscule, sap-siphoning, genetically identical daughters. Aphids might lead us to ask not "why do they clone?" but "why don't we?" Shouldn't the long-term female homosexual pair bonding in certain species of gulls help define our views of successful parenting, and help [us] reflect on the intersection of social norms and biology?[81]

The instability of sex spans species. Books such as Joan Roughgarden's *Evolution's Rainbow* and Bruce Bagemihl's *Biological Exuberance* chronicle the immense variation of queer animal sexuality. These works and others seek to challenge the naturalization of heterosexuality.

QUEER BIOLOGY

In a provocation that has stayed with me, Elizabeth Wilson, while discussing Charles Darwin and his work on barnacles, poses a critical question on theorizing sexuality. Though barnacles are usually hermaphrodites, in 1849 Darwin discovered a species where males become parasitic within the sack of the female (what he called "complemental" males). In another species he observed a hermaphrodite with embedded complemental males, a species he speculated as halfway between a hermaphrodite form and a sexed one. In yet another, he found a hermaphrodite with over seven complemental males attached. Darwin thought he had uncovered transition points from a hermaphroditic form to a bi-sexed form—a multitude of genders and sexes. They represent intermediary stages of sexual diversity *and* somatic diversity. In considering Darwin's discovery, Wilson insightfully notes:

> To characterise Darwin's barnacles as queer is too glib—if by this characterisation we mean that the barnacle simply mimics those human, cultural and social forms now routinely marked queer (the transgender barnacle! the polyandrous barnacle!). This characterization has more punch if it is used, contrariwise, to render those familiar human, cultural and social forms more curious as a result of their affiliation with barnacle organisa-

> tion. The queerness of Darwin's barnacles is salutary not because it renders the barnacle knowable through its association with familiar human forms, but because it renders the human, cultural and social guises of queer less familiar and more captivated by natural and biological forces.[82]

How can we renarrate the world so that the human and its social, sexual, and racial organization is not centered? How do we recover and celebrate the diversity of cultures and sexuality? How can the human be rendered anew, as a tiny fraction of the biological splendor of the planet rather than an exceptional or grounding figure?

Science has developed a critical reliance on numbers, data, and statistics as representations of nature. If the dead herbarium sheet was the mark of empire, then contemporary biological research and its reliance of numbers, in the guise of objectivity and rigor, is even more disembodied. I am returning to traditions in plant biology and indigenous cultures that focus on a different mode of storytelling—plants as embrangled planetary coinhabitants, as potential kin in evolutionary time.

Fables for the Mis-Anthropocene: The Queer Vegennials

From the Chirp-Net emerged the power of the Queer Vegennials, an inventive and innovative transnational group, a generation of young decolonial plant lovers whose mission is to undo the colonial gaze. Our understanding of plants has been dominated by humans, literally through forestry, domestication, agriculture, and plantations, but also through theories of plant biology. The Queer Vegennials (QVs) recognize and appreciate the astonishing diversity of plants, from mosses and ferns to flowering trees. They are resolute: plants defy every rule! QVs celebrate the many exceptions. Tumbleweeds tumble across the country. Plants lure insects, birds, mammals, and the elements to transport pollen and seeds across vast geographies. Cacti don't have traditional leaves. Flytraps and pitchers trap insects. Paintbrush and dodders do not photosynthesize. Most bromeliads don't need soil. Epiphytes trap water and nutrients from the air. The beloved mistletoe is parasitic. Plants engage in mutualistic relationships with bacteria and fungi. Exhibiting immense variation, no two plants or flower are the same.

Dynamic and mobile, plants radiate into the sun to harness solar energy and root deep into the ground for water and nutrients. They engage a diverse soil community below and land community above; their pollen and seeds travel great distances hooked to hair, feather, and fur or riding the currents of air and water. Plants are immersed in the world around them.

Refusing the binary male and female, QVs reverted to old terminology—pistils, stamens, pollen, ovules, and seeds. *Use more words*, they said.[1] Plant sex is a profound act of sociality. QVs insist there is no one way to be social *or* sexual. Plants provide nectar, milk, food, and homes for many. Why reduce plants to humans? Why not embellish and inspire human capabilities by using plants as models? Plants are dynamic conglomerations of multispecies and elemental engagements. Based on this simple premise, QVs produced new models of humanity inspired by plants. They initiated the Metaphor Project. The philosophy behind the Metaphor Project was not to claim a biological basis for behavior or organization but to recognize that the natural world could inspire. But, as their charter states, the natural world can also be very violent and harsh. The project drew only on models that fostered a queer politics.

QVs from South Asia initiated the Banyan Project. The banyan, a fig tree, produces aerial roots that grow toward the ground, take root, and grow into new trees. A single banyan tree can produce a whole grove. The Banyan Project served as a metaphor for the Chirp-Net—a cooperative where new branches are self-sufficient, a nonhierarchical, directionless infrastructure capable of continuous growing. In India, banyan trees symbolize permanence. It was under a banyan tree that the Buddha achieved enlightenment. The Banyan Project transformed the world. New groups emerged with similar models based on underground runners and rhizomes, tubers, corms, and bulbs. The Chirp-Net was well networked!

Perhaps QVs' most successful initiative was the Dungowan Project from Australia. It drew on the bush tomato, a plant with no stable sexual expression, as its metaphor. The Dungowan bush tomato, with its fluid breeding system, has any number of fully staminate, pistillate, and bisexual flowers. It has every form of flower and mode of reproduction imaginable. The project inspired the pansexual movement and a radical agenda. In an era when reproduction and sex were decoupled, endless possibilities of social and sexual intercourse and formations of kin were possible. Social communities with multiple parents of multiple sexes and genders embraced a plethora of sexes, genders, sexualities, and socialities.

QVs in the tropics (and some from elsewhere) began the Orchid Project. They rewrote traditional narratives of orchids. They celebrated orchids and their intimate histories of coevolution of plant and animal. The project challenged the pejorative readings of orchids as false and deceitful, where the duplicitous orchid "cheats" the bee. Through a series of experiments, QVs challenged the notion that pleasure and sociality are exclusively animal traits. It refused the mathematical and "game" calculus that drove much of evolutionary biology. In challenging the rigid borders around sexuality and sociality, the Orchid Project rewrote plant reproductive biology to celebrate plant and insect sensuality, joy, and play.

QVs from the Arctic began the anarchic Beanion Project. Growing amid harsh conditions that supported little other life, beanions were marked by unusual phenotypic plasticity, an incredible capacity to transform into sexual and social modes depending on environmental contexts. It was impossible to attach models of pleasure, desire, or sexuality to this species. It defied logics. For many, the Beanion Project provided solace from worlds that were fixated on desire, pleasure, and productivity. "Be yourself" was the motto. That's enough.

QVs from the Americas introduced the Cuckoo Project. Jack-in-the-Pulpit, or the cuckoo plant, looks like a pitcher plant but is not carnivorous. Like the orchid, botanists malign the conniving plant that lures insects into its pitcher and then shuts the lid for a while. The trapped insect, portrayed by scientists as disoriented, confused, and panicked, flies around the pitcher covering itself with pollen, and when released it facilitates cross-pollination. The Cuckoo Project, through a series of experiments, showed that insects enjoyed the playful merry-go-round, often seeking such plants out. Such long-term coevolution was mutually enjoyable. These QVs also challenged tired old stories of "battle of the sexes" that pitted males and females against each other, and cross-species engagements in the language of competitive battles. Instead, they encouraged new rhetorics of mutual alliance, friendship, mutualisms, pleasure, desire, and play, language long marginalized in evolutionary biology.

Pan-African QVs introduced the Baobab Project. The baobab, an ancient flowering plant, has a water-storing trunk that can provide food, water, shelter, and medicine to humans and nonhuman animals. Its bark produces strong fibers for rope and cloth. The project inspired communal visions of culturally and religiously important plants creating a thriving multispecies community of mutual dependence. With the transformations of global ecologies, QV communities modeled thoroughly entangled ecological communities.

A global QV initiative, the Slime Mold Project, emerged as an even more radical project. Eukaryotic organisms, slime molds live freely as single cells, but they also aggregate together to form multicellular bodies and collectivities, challenging the very idea of individuals. This project inspired a dynamic and supple organization, with radical and decentralized models of decision making, opening up models of collectivity and community.

The most interventionist of the global QVs was the Bacteria Project, which ushered in a world of radical kinship. They argued that while much of human biotechnology was stolen from innovations of bacteria and viruses, technologists embraced the wrong lessons. Rather than short-term logics of biotechnology and its technological determinism, the Bacteria Project celebrated bacteria as key mutualists in the complex web of life. Every multicellular organism is a mutualism. Bacteria and viruses have moved genes across species for millennia, helping us see that evolution is both vertical and horizontal, across generations and species. In moving genes across species, they bypass evolutionary forces to enable huge jumps in evolutionary possibilities. In the end, bacteria help us see that we are all related. Humans should discard ideas of kinship modeled around patriarchal notions of the family, they argued, and embrace the radical kinship enabled through bacteria and viruses. After all, for every human "cell" there are ten times as many bacterial cells. Onward the Raucous Spring!

INTERLUDE

International Council for Queer Planetarity: The Botanical Debates

What if a community of queer botanists, the "we" in this story, assembled to address the ravages *of* climate change? Entirely a work of fiction, this piece draws on queer studies to rethink biological knowledge.

We are decades into the ravages of climate change. There are pockets of good news—a slowing here, a reversal there—but we are, without question, on a steady path to increasing temperatures, dangerous CO_2 levels, deadly pollution, extinctions, ecosystem devastation, erratic weather, droughts, floods, fires, and depleted water tables worldwide. The planet is growing inhospitable. In response, the wealthy have built weather bubbles to filter clean air and water into their homes. The planet's marginalized human and more-than-human worlds yet again bear the brunt.

The council has discussed and debated the state of the planet. We trace the problems to the enduring extractive logics of colonialism, conquest, and slavery. Colonialism endures in settler colonialism. Colonized nations "underdeveloped" the colonies by stealing their resources with no recompense. Resources continue to be exploited by elites. Short-term benefits and profits are consistently prioritized over planetary health.

Action is urgently needed. We needed a new countercolonial blueprint.

MANIFESTO FOR QUEER PLANETARY LIFE

One first act was to revisit the membership of the council. We had to confront our colonial, racist, sexist, and ableist history. History has shaped us at every level. Science and technology are now in the throes of elite power. We reached out to communities that live with the land—indigenous, pastoral, tribal, land-based, and queer ecological communities long marginalized by an increasingly scientized environmentalism.[1] We began by decolonizing the membership. Reparations and repatriation were key. Whenever possible, we returned land to the stewardship—not ownership—of communities that had sustained relationship with the land.[2] We offered our resources and labor to assist them when they called upon us.

There is no singular indigenous. We will not fetishize and ossify living and dynamic cultures. Histories of settler colonialism and ongoing repressive practices have taken their toll. Patriarchy, violence, and food and health insecurities have decimated many indigenous communities. Yet many have persisted, indeed thrived, nurturing and building engaged relationships with the land and its more-than-human inhabitants. Their treasured knowledge is critical for the planet.[3] We recognize these many histories and strive to develop new modes of engagement that own up to the unequal relationships of knowledge and power. Conversations to arrive at our various decisions were often troubled, contentious, and complex. By constantly recreating and revisiting rules, we moved forward.

It has become clear that we must prioritize nonhierarchical thinking (often nonwestern, indigenous, and marginalized western). We eschew the hierarchical thinking of the "great chain of being," a concept that arose in medieval Christianity.[4] Scientific theories have often been hijacked into theories of progress, human superiority, eugenics, and environmentalism. A potent and enduring politics of identity continues to undergird much of scientific thinking.[5] Our fundamental goal now is to challenge. We move away from scientific objectivity to engage with the world in all its complexities. We are *in* and *of* this world and pledge to work within and with it. We welcome other cosmologies of interconnected and entangled ecologies. Plants are the primary autotrophs (along with some algae and bacteria).[6] Life on earth depends on them. Our immediate focus is to celebrate land and soil and prioritize photosynthetic organisms (PSOs). All life depends on the success of PSOs.

The council once was populated by elites. Our first task, then, was to rectify ideologies of dominance and control and claims of manifest destiny. Through long and at times tortured dialogue, we convened a representative body that does not privilege elite power.

This year, we the Council for Queer Planetary[7] will focus on photosynthetic organisms (PSOs), or greening our planet.[8] We will immediately begin with decontaminating the air, dealing with escalating levels of CO_2 and our depleting ozone levels. Once we achieve minimal success, we will expand our goals.[9]

One issue has loomed large—scientific histories of sex, gender, and sexuality. Our foundational theories of sexuality ground the plant sciences. In promoting PSOs, we must promote a diversity of adaptations—omni, uni, trans, pan, ambi, sexual, multi, poly, co, and inter sexes and genders, mixed mating systems, as well as those who self-fertilize or inbreed.[10] In expanding our understandings of the world, we are attuned to critiques of scientific theories of sexuality. To that end, we believe:

- We must not privilege the sexual over the social. There is much more to biological life on earth than a singular focus on sex. Our work focuses on detailed and engaged observation to understand biologies and ecologies in all their complexities.
- Science has erroneously characterized the natural world as a binary system of sexes (male/female), genders (masculine/feminine), and a complementary sexuality (heterosexuality) as the natural or normal state. A whole edifice of biology is built from this false assumption.[11] But in biology, there are no rules. If you are convinced sex is binary, look at plants. If you believe in the biological species concept, look at the many hybrids. If you believe in individuals, look at fungi. If you believe sexual reproduction is the key to life, look at species who have survived millennia without. If you believe that sexual reproduction and vertical inheritance are the key to evolution, look at bacteria and viruses that move genetic material across species. Zoonotic viruses jump from animals to other types of animals, including humans. Phytonotic viruses do the same from plants to other organisms. Life is dynamic and hypermobile.
- The goal cannot be stability but must be engaged living.
- Modern biotechnology learned its techniques from the queer brilliance of bacteria, a world teeming with difference—an expanse of promiscuous perversity! We must use it ethically.
- We need to rethink foundational ideas of individuals, cells, and cell or-

ganelles. We believe that mitochondria, the powerhouse of the cell, were once independent organisms that entered into a symbiotic relationship with another cell to first evolve multicellularity.[12] Evolutionarily speaking, sexual reproduction is a recent phenomenon. Some scientists argue that sexual reproduction evolved by accident as a necessary byproduct of the evolution of multicellularity and cellular differentiation.[13] In short, ideas of binary sex and sexuality are not an inevitable part of settled theory.

- Our cells house multiple organisms, and most animals host more bacterial cells than their own; we are individuals but holobionts.[14] Bacterial diversity is critical to planetary life.[15] The use of antibiotics and the rhetoric of war and aggression against microbes have seriously harmed the planet. We eschew the language of battles and wars as models of multispecies life. Instead, we embrace engaged living.
- Sex and sexuality are far more complex than their binary formulation in biology. In expanding categories and vocabularies of sexuality, we need to do the same for sex. Not all heterosexual pairs can reproduce because of evolved genetic compatibilities and incompatibilities. These are starting points to think in terms of multiple sexes, genders, and sexualities that create complex, stratified, and reticulated evolutions. There is more to sexual and social life than a biology grounded on assumptions of heterosexuality.
- Evolutionary theories are grounded in assumptions of randomly mating populations, yet human societies are both segregated and stratified.
- We must address the violence that humans have wrought through forestry, agriculture, and the plant sciences.
- Human societies across the world have reclaimed more complex models of sex and sexualities. An explosion of global queer movements celebrate sexual diversity. We applaud and celebrate knowledges long forgotten. How do we bring their insights into biology?
- We have refashioned our language of plants. Rather than anthropomorphize them, we have returned to earlier terminology. We retain the words *pistil* (like a pestle), *stamen* (threadlike), *pollen* (fine powder), and *seeds* (to sow)—affective and functional terms that describe the morphology of the plants. This is a step toward new words and worlds.
- Binary sex is premised on anisogamy, the presence of two distinct gam-

etes of varying size. All other cells in the bodies of multicellular organisms are intersexed. We must stop sexing the body, dividing creatures into males and females, when the vast majority of cells show no sexual differentiation.

- When we select flowering plants, should we privilege species with hermaphroditic plants, in particular bisexual, perfect, and complete flowers (these are botanical terms—"perfect" flowers have pistils and stamens, and "complete" flowers have sepals, petals, pistils, and stamens)? No! We are troubled by the very language of perfection.
- All organisms on earth have adapted to their environments. They make informed decisions to flower, drop leaves, and abort fruit for strategic reasons—adaptations to their evolutionary histories.
- Do we care about modes of reproduction? Yes and no; diversity is key. We need diverse sexualities.
- Working with local communities, we assist some species into more habitable environments. There is no nostalgic return to the past but an imaginative move to the future. We have created many such sites across the world.
- In short, our very queer world invites us to rejoice and learn to live with joy, engagement, and playfulness.

We have initiated an international compendium of lost and forgotten ideas—an ecological and intellectual "planetarium," a collection of local words, concepts, and epistemologies long forgotten or erased. Words are powerful; we need more words to adequately capture the enormous diversity of life on earth.

How do we re-green the planet? Do we just open up spaces and hope for a return of plants? No![16] In choosing organisms, should we be randomly representative? After some deliberation, again no.[17] We believe that our actions should be deliberative, thoughtful, and respectful to all creatures. We must work with communities who understand biologies and ecologies. Given the variation in habitats and environments, some creatures are better adapted and better equipped to heal, nurture, and recuperate the planet. But many questions remain: Which regions of the world should we focus on? Who is best adapted to the current and future environment? Which creatures can withstand the erratic pattern of environmental changes? Here we draw on evidence from history.[18] We use scientific projections to imagine possible futures and select organisms that will thrive in a fast-changing planet.

Organisms are interdependent and often need each other for resources such as food, pollination, and dispersal. We must work to sustain and nurture *habitats* and *communities*. We are particularly mindful that soil communities are critical to photosynthetic organisms—many have coevolved complex relations with other soil bacteria, endo- and ecto-mycorrhizal fungi, yeasts, myxomycetes, actinomycetes, and nematodes. Symbiotic organisms have evolved over millennia, and organisms are dependent on them—bacteria fix nitrogen, and fungi provide water and important nutrients. Equally important are parasitic and pathogenic organisms, as well as saprophytic organisms that help in decomposition to produce resources for life on earth. Life is a complex web; let us not label its parts with human-centric ideas of "good" and "evil."

Rather than focus on traditional biological classification, we embrace the elements—air, earth, water, and fire. Earth, rock, and soil are not inert backdrops. There is a cycle of life. After all, coal in the mine was once a tree. Elemental forces provide critical nutrients, disperse spore, pollen, and seeds, nourish the earth and nurture its inhabitants, and regenerate life by circulating carbon. Creation and destruction, life and death, are part of the cycle of life. We must create dynamic and diverse systems, not monocrops and monocultures. Colonial environmental management destroyed the planet by creating droughts and floods, uncontrolled fire, hurricanes, and tornadoes. We must nurture the planet's regenerative capacities—the understory and soil communities as critical to life as the magnificent overstories.

We may get it wrong, but we must *try* with care and a sense of responsibility. Technology is important, but we must refuse a mindset of technological determinism. Using all our senses, we must deliberate with history as our guide.

Are invasive species worrisome? No; they can be our saviors. We do not want monocultures, but the category of invasive species is not useful.[19] We need to foster all kinds of plants: overstories, understories, and midstory plants. Communities need organisms that are short-, mid- and long-lived.

There is a temptation to choose organisms that reproduce rapidly (r-selected) as opposed to more slowly (k-selected). In many damaged landscapes, r-selected species are indeed critical. However, to sustain life, we need a diversity of life histories.

Finally, we have initiated new educational curricula with these key insights. Students should not have to study knowledge only to have to unlearn it later. Tools to live in an entangled world are lessons we need to refresh each day, starting as young as possible.

There is pleasure in the world—sociality, joy, desire, play. And so, we hope, we will build a world in collaboration, in community. Life and death, selfishness and sacrifice, greed and kindness always persist. So too do symbiosis, symbiogenesis, and the grand mating of bodily exchanges and mixing of genetic material fueled by bacteria and viruses, the biochemists of nature. Life forms will change, adapt, readapt. There is always drama and novelty, never the same old story. This is the grand life on Earth.

PART FOUR

Pangaean Dreams

MAPPING BIOGEOGRAPHY

Perhaps you have forgotten. That's one of the great problems of our modern world, you know. Forgetting. The victim never forgets. Ask an Irishman what the English did to him in 1920 and he'll tell you the day of the month and the time and the name of every man they killed. Ask an Iranian what the English did to him in 1953 and he'll tell you. His child will tell you. His grandchild will tell you. And when he has one, his great-grandchild will tell you too. But ask an Englishman—" He flung up his hands in mock ignorance. "If he ever knew, he has forgotten. 'Move on!' you tell us. 'Move on! Forget what we've done to you. Tomorrow's another day!' But it isn't, Mr. Brue." He still had Brue's hand. "Tomorrow was created yesterday, you see. That is the point I was making to you. And by the day before yesterday, too. To ignore history is to ignore the wolf at the door.

JOHN LE CARRÉ | *A Most Wanted Man*

CHAPTER SEVEN

Botanical Amnesia

Colonial Hauntings in Plant Biogeography

Land is very important to Indigenous Peoples, but we think of land quite differently from the colonizers. For us, land is not an enclosure that is protected by a border. Land is not a natural resource to exploit. Land is not a commodity. It is a particular space full of relationality to which we form very deep attachments over very long periods of time. . . . If you have a world where relationships and process are paramount, land and bodies become so very connected in relation to both space and time.

LEANNE SIMPSON | *Temporary Spaces of Joy and Freedom*

Before you ask other people to respect the borders of the West, ask yourself if the West has ever respected anybody else's border.

SUKETU MEHTA | *This Land Is Our Land*

I'M LEARNING TO TELL STORIES IN PLANT TIME, in planetary time. This is how it goes. Until about two hundred million years ago, Eurasia and the Americas were part of a single landmass, Pangaea. The supercontinent fractured into tectonic plates that floated into new continental formations, into places and spaces filled with new species and new collections of kith and kin. The plates continued to move, carrying species to novel environments, leading to a dizzying diversity of adaptations, innovations, and imaginations. Then, in a flash of planetary time, colonialism reknit the "seams of Pangaea."[1] Colonists traveled, conquered new lands, imposed new regimes of life and living, and moved species around the world willy-nilly. Colonialism was fundamentally an ecological project, affecting not only humans but all planetary life. Ever since 1492, parts of the world have grown more alike as ecosystems have collided and intermingled. Alfred Crosby's groundbreaking work in *The Columbian Exchange* documents how the biological exchange and admixture brought by colonialism created a world-spanning economic system where plants, animals, and people shuffled across the globe. Bananas and coffee, two African crops, are now the principal agricultural exports of Central America; we would not have the cuisines of modern Thailand, Italy, and India without peppers and tomatoes form Mesoamerica. The Columbian Exchange is the "reason there are tomatoes in Italy, oranges in the United States, chocolates in Switzerland and chili peppers in Thailand. To ecologists, the Columbian Exchange is arguably the most important event since the death of the dinosaurs."[2] As continents drift into a New Pangaea, what stories await us?[3] It depends on how we act now.

I revisit the idea of invasive species through the long arc of planetary time, centering the violence of colonialism. Biogeography is a field concerned with the distribution of species and ecosystems in geographic places and through geological time. A focus on invasion biology gets to the heart of how biologists narrate and conceptualize questions of place and time. Place and time are also rich territory for the humanities and the social sciences. As a result, these concepts offer rich terrain for interdisciplinary thinking.

Over the last two decades, I have tracked the evolving discourse on invasive species. I have watched the growing xenophobia around human immigration transfer onto plants and animal worlds.[4] I witnessed the heightened vitriol of violence against foreign humans, animals, and plants after the attacks on the Twin Towers on September 11, 2001. If US Homeland Security instituted regimes of surveillance of the foreign, environmental organizations matched these in botanical infrastructures. Some of the effects of xenophobia are personal. During the pandemic, as I quietly gardened in my yard, a man walking by thanked me for the virus. For immigrants, xenophobia is everywhere. And just as the country wrestles with the politics of immigration, biologists wrestle with the language of invasive species.

If the world is indeed naturecultural, should vocabularies be transferred across natures and cultures? Yes and no. A careful naturecultural analysis is critical. For example, I believe the xenophobic responses to foreign humans, plants, and animals are related. A robust literature supports this conviction.[5] Our anxieties about social incorporation (associated with expanding markets, increasingly permeable borders and boundaries, growing affordability of travel, and mass immigration) have historically spilled into our conceptions of nature. Nancy Tomes, for example, documents how panics about germs in the United States have historically coincided with periods of heavy immigration of groups perceived as "alien" and difficult to assimilate.[6] She documents the germ panics in the early twentieth century in response to the new immigration from Eastern and Southern Europe, and in the late twentieth century to the new immigration from Asia, Africa, and Latin America. "Fear of racial impurities and suspicions of immigration hygiene practices are common elements of both periods," she writes. Like these early germ panics, questions of hygiene and disease haunt exotic plants and animals. The long arc of planetary time reminds us that xenophobic rhetoric emerges periodically to engage nationalisms and nationalist politics. We live as collectives, albeit hierarchically.

Are Europeans invasive species? I would suggest not. Foreign plants move, sometimes due to colonial and capitalist hubris, and largely because of human actions. Unlike colonizers, they do not send spoils from their new lands to enrich home populations. Foreign plant movement and human colonialism are not analogous phenomena. Likewise, plants are not colonizers. Calling them so depoliticizes the profound violence and extractive regimes of colonialism. How then do we contextualize the power and violence of colonialism? In this chapter and the next, I explore this question through the vantage points of settler colonialism and postcolonialism. They yield very different answers.

Colonial histories and ongoing capitalist greed have devasted lands and ecosystems. Damaged ecosystems have created the grounds for some (native and foreign) plants to thrive. For example, the spread of kudzu in the United States and water hyacinth in India are real concerns for many reasons. But should we call them by pejorative terms such as "invasive" species? Should we develop a litmus test of belonging by labeling plants as native or foreign? This chapter and the next outline the inaccuracies behind the stories we tell about invasive species, and the violence that results from our carelessness.

Rather than foster xenophobia, we should challenge *native* and *foreign* as labels for humans and plants. We need new vocabularies and new stories to narrate life on earth.

Race and the Coloniality of Power

Central to the colonial project was the invention of the scientific concept of race, a cornerstone of the coloniality of power. To exploit the colonies, the Spanish empire developed complex and hierarchical systems of racial categories, leading to contemporary racial classifications of biological inferiority and superiority.[7] The idea of race, Quijano argues, is "the most efficient instrument of social domination invention in the last 500 years."[8] The infrastructures of coloniality aided the efficient exploitation of laboring bodies and natural resources. In short, coloniality is a constitutive component of modernity and not a derivative feature; "there is no modernity without coloniality."[9]

The history of "race" is central to biology. In plant nomenclature the term *subspecies* or *race* captures variation between individuals. Linnaeus used the term *varieties*; many view Linnaean varieties as subspecies or "races." However, for Linnaeus, human races emerged from divergent geographies, and it was geographical contexts that endowed different groups with varied physical, moral, and intellectual capabilities.[10] While "race" largely remains a category of the human, it was once a category for all organisms. Sylvia Wynter traces the figure of Man to secular reformulation of a "Judeo-Christian Grand Narrative."[11] The racialized othering was biologized in racist anatomical characterizations.[12] And geographies of difference transcend into plant worlds.

Botanical Amnesia: Empire's Convenient Memory Lapses

The emergence of the field of invasion biology is most often credited to Charles Elton and his influential book *The Ecology of Invasions by Animals and Plants*, first published in 1958.[13] In this book, he lays out the case for why we live in a "very explosive world." He uses the word *explosion* more than *invasion*—perhaps the new science might have been called "explosion ecology" instead of "invasion biology"![14] As Elton frames the argument:

> It is not just nuclear bombs and wars that threaten us, though these rank very high on the list at the moment: there are other sorts of explosions, and this book is about ecological explosions. An ecological explosion means the enormous increase in numbers of some kind of living organism. . . . I

use the word "explosion" deliberately, because it means the bursting out from control of forces that were previously held in restraint by other forces.

In this book, Elton introduces a new field invasion biology, but the field did not blossom into its configuration until the 1980s.[15] To begin with, what is an invasive species? The US Department of Agriculture (USDA) defines invasive species as species that are nonnative (exotic/alien) and cause harm to the economy, environment, and human health.[16] Since the 1980s, the field of invasion biology has exploded.[17] In recent years, a frenzy of xenophobic alarm by groups from the political right and left, from environmentalists and nonenvironmentalists alike, have spawned a veritable industry of bioinvasion: an entire academic, policy-oriented, and activist field. It has proved a fertile ground for large investments of money, attention, and resources into research, policy, and activism: the National Science Foundation has created committees and grants; a plethora of scientific and popular publications are devoted to bioinvasion; international, national, state, and local governments have implemented policies; and environmental groups such as Nature Conservancy, Sierra Club, and Greenpeace have all developed robust programs in invasion biology.

The idea of "invasion biology" is predicated on a view of "nature in place" and "nature out of place." This growing international interest in the field has stoked alarm about a world increasingly "out of place." Cultural theorists argue that this recent hyperbole about alien species parallels germ panics of the past and is best understood as a cultural panic about changing racial, economic, and gender norms in the nation. The perceived globalization of markets, and the real and perceived lack of local control, feeds nationalist discourse. These shifts continue to be interpreted by elements of both the right and the left as a problem of immigration. Thus, immigrants and foreigners, products of the global, are perceived as one of the key reasons for problems in the local.

Thinking of invasion biology through the histories of empire is enlightening because it reveals a profound botanical amnesia. Our theories of nature often do not consider the role of empire in shaping the biogeography of nonhuman entities in the world. Yet empire is as central to nonhuman worlds as to human worlds. Where humans went, plants, animals, fungi, bacteria, and viruses went. Environmental historians have argued that we should understand imperialism as *fundamentally an ecological project* in which humans, plants, and other species were shuffled around the earth in schemes for colonization and conservation. In the early modern period, botany was big science and big business, critical to Europe's ambition as a colonial trader. Colonialism ushered in a grand reshuffling of global biota, a biological bedlam that reknit the seams of Pangaea.[18] Indeed, it would be accurate to characterize colonial expansion as the original bioinvasion. One can and should understand the botanical sciences as a significant legacy in the afterlives of empire. Colonialism fueled an extractive economy through the objectification and commodification of the colonized world and the destruction of local knowledges. In its place, the colonial order installed the biological sciences as the singular universal, abstract, and expert knowledge. Local ecologies were transformed by colonial logics. Colonial movements of plants included the famous spice routes, extracted lumber resources, introduced plantation crops, and global transports of horticultural specialties and a trove of other agricultural plants and animals. As a result, in most countries, agriculture includes predominantly foreign plants and animals. Horticultural societies and gardens across the West cultivated the exotic and curious from around the world; Kew Gardens and other such sites became repositories of the world's biota.

Thinking of contemporary invasion biology alongside colonial botany brings into stark focus the hubris of colonial logics. During colonial rule there was a laissez-faire attitude to unlimited mobility across borders. For example, in the late nineteenth and early twentieth centuries, the USDA had an active program where biologists as "explorers" roamed the globe in search of new and interesting plants of economic and aesthetic interest. Dr. David Fairchild, director of USDA's Section of Seed and Plant Introduction from 1898 to 1928, is said to have personally introduced more than two hundred thousand species and varieties into the United States.[19] Likewise, the American Acclimatization Society introduced a variety of plants and animals and attempted to introduce all of the bird species mentioned in Shakespeare's works to New York City's Central Park in the 1890s—including the sparrow and the starling, two reviled bird species today.[20] Such an openness was the norm until late in the nineteenth century. Further, as Alfred Crosby argues, the roots of European domination of the western world lie in their creating "New Europes" wherever they went; settler colonists ravaged native populations of humans, plants, and animals.[21] Where Europeans went, their agriculture and animals went; they thrived while indigenous ecosystems collapsed. The mass export of resources out of the colonies, and the growing ravages of the environment through unchecked industrialization and logics of development, shaped the afterlives of empire through neocolonial policies and programs.

The arrival of foreign plants and animals is deeply intertwined with human desires for food, ornamentals, recreation, utility, comfort, entertainment, soil erosion control, and pest or weed control.[22] In contrast to the invasive species paradigm, postcolonial studies and diaspora studies highlight the naturecultural consequence of colonialism and its regimes: the (forced) migrations of people and plants. Drawing on Arturo Escobar's theory on "imperial globality,"[23] Ogden astutely observes that what are labeled as invasive species are plant populations "whose mobility is predicated on their incorporation into economic projects that have global configurations."[24] Anti–invasive species activists never work to exterminate agricultural species, the majority of which are foreign. As long as exotic or alien plants "know their rightful place as workers, laborers, and providers, and controlled commodities, their positions manipulated and controlled by the natives, their presence is tolerated."[25]

The vast majority of invasive species arrived because powerful humans invited them and facilitated their entry; most biological invasions are *invited* invasions.[26] Histories of race are woven into our conceptions of invasive species.[27] In the United States, the Asian carp was imported as a "worker fish" to clean up areas with aquatic weeds, and kudzu was promoted by the government to prevent soil erosion.[28] Let's consider kudzu more closely. This beautiful perennial vine in the pea family grows throughout Japan rather innocuously and is used there as food, medicine, animal fodder, and fiber for weaving.[29] The plant, first exhibited in the United States during the first World's Fair in 1876, was an instant hit. It had innumerable desirable properties. It could be grown on any soil, endure any weather, and as a legume, it could fix nitrogen into the soil, thus nourishing it—an ideal candidate to prevent erosion. In the 1940s, the American Civilian Conservation Corps hired hundreds of volunteers to plant kudzu across the American South; farmers were paid as much as eight dollars an acre to cultivate it. Kudzu thrived in the South, better than in Japan—and indeed better than anyone imagined. Today, it is portrayed as a foreign bioweapon choking out native species: "the vine that ate the South," "mile-a-minute plant," "foot-a-night vine." We must ask: Is the villain here kudzu, or human hubris? Labeling the plant as evil is practicing botanical amnesia.

More than half of the plants known to be invasive in North America were originally imported for their horticultural use.[30] These invasions proliferated through domestication, acclimatization, and breeding sciences—scientific disciplines that converted "foreign" species into economically valuable ones.[31] In the same spirit, botanical gardens introduced citizens, especially children, to many species not native to the United States, ushering in a long history of fascination with exotic plants from around the globe.[32] Human dispersal is key for the movement of plants and animals.[33]

In the context of the history of empire, the rise of invasive species discourse is particularly ironic. Even as colonialism had moved the world's biota and ushered in new landscapes of empire and new formations of naturecultures, with the advent of freedom in the postcolonies, nativist thinking reengaged the national borders in the west. Indeed, discourses of invasive species remain a prominent feature of the environmental afterlives of empire, as waves of nationalisms have consumed the postcolonial worlds and colonial nations have moved to secure their borders. In an ultimate act of botanical amnesia, after centuries of global expansion and ecological decimation, those very nations now insist on preserving the "new Europes" from newer immigrants from Asia, South America, and Africa.[34] Tracking the histories of "invasive" rhetoric allows us to understand how plant and animal quarantine laws developed alongside national exclusion acts. Indeed, in the poignant cases of settler colonialism, national rewriting of history indigenizes the white settlers as the new "natives" that now relegate all others as foreign and undesirable.

I renarrate this abbreviated history to highlight both the hubris of empire and its subsequent amnesia, which together prevent a true botanical reckoning with our colonial past. It is a very convenient amnesia. Having unleashed the biological bedlam that colonialism wrought on the natural world, western nations now beat the thunderous drums of nationalism to keep their stolen spoils of conquest. As Suketu Mehta's epigraph to this chapter wonderfully summarizes, the west never respected other nations' borders, and yet through quarantine and immigration laws it now deems its borders sacrosanct and polices them through increasingly militarized zones. Western countries consistently deploy dehumanizing languages when describing immigrants, a continuation of the "civilizing" mission of colonialism. Seen through this long arc of history, invasion biology emerges as entirely a project of colonial consolidation, nostalgia, and return. These countries want to secure the spoils of empire and also want to return to a moment in colonial history when white nations reigned supreme, a time when the colonized knew their place as colonial subjects.

Thinking about colonialism and environmentalism together, the nostalgia that accompanies them both seems futile. A botanical reckoning is a profound recognition that frameworks of "nature in place" and "nature out of place" are political constructions, attempting to fossilize particular ecological/political moments as "true native nature" for posterity. Understanding the world as dynamic naturecultures reminds us that despite our cultural propensity for nostalgia, the biota of the world have their own agency and propel their own evolutionary futures. Biological nostalgia is a fundamentally unecological project. We do not, and cannot, control the world, even if we wished to.

Colonial Hauntings in Biogeography

Colonization haunts plant biology in profound ways. While plants are never consciously part of an ethnonational group, colonial biology often assigns geographic monikers to plants and animals, not to mention officially and scientifically naming them after nations and famous colonists. In an anthropocentrism that pervades botany, plants are assigned ethnic and national categories such as the Asian long-horned beetle, Asian carp, Brazilian pepper, Chinese tallow, Cuban laurel fig, English ivy, European starling, Florida holly, small Indian mongoose, Japanese barberry, Japanese honeysuckle, Japanese knotweed, Java plum, and Norway maple. Does the tallow or maple know the borders of the nation? Such nomenclatures fossilize organisms into prized origin stories, and indeed settler-colonial origin stories. The very naming of fields like "restoration biology" evoke a turn to the past, a settler version of nostalgia. As always, those at the center of power dictate the naming even while they remain nameless, without geographies. As Toni Morrison astutely notes, "In this country American means white. Everybody else has to hyphenate." Settler logics unfold into a poignant irony where native peoples are rendered foreign, never true Americans.

Over the last two decades, I have, on numerous occasions, presented my work on the problematic framing and usefulness of invasive species as a biological concept. All too often, biologists moved by the argument approach me and say something like, "I get it. We should move away from the language of invasion. What word should I use instead? What term will avoid these racist histories?" I sigh and repeat again that my objections aren't about a word but about the conceptual universe of botany and its geographical litmus tests that divide the world into binary categories of *good native* and *bad foreigner*. This framing leads to reactionary politics—for biology, the environment, and cultural worlds.

If we believe that some plants are overrepresented, there are other ways we can address the ecological imbalance. For example, rather than a geographic litmus test of native/alien, biologists have suggested focusing on other life history traits—dispersal distance, uniqueness to a region, or impact on a new environment—as ways to identify which small fraction of species might potentially become a problem.[35] We could locate our policies around biology, not colonial politics. Imagine that! Why categorically profile all foreign species as evil and undesirable?[36]

Many such alternates have been proposed by biologists. Yet the rhetoric of invasion biology endures to channel the virulent immigration policies and politics that continue to inflame. The solution is always a xenophobic and often violent one, to eradicate foreign species: poison the lake, burn the forest, dump chemicals, spray poisons. In addition to these large-scale efforts, people are mobilized to eradicate individual invasive species. These efforts evoke a rememory of colonial haunting—the annihilation and eradication of the worthless "other" and the fantasy of righteous and entitled management. It appears easier to scapegoat foreigners than to interrogate the proximate causes of invasion.

Most species considered invasive in a place today actually arrived decades if not hundreds of years ago.[37] The question we should be asking is why they are turning invasive now. What has changed? Environmental degradation? Development policies? Overdevelopment? Widespread disturbance? Pollution? Climate change? Despite plenty of evidence of the problems caused by disturbance through overdevelopment of land,[38] much more effort goes into eradicating species than challenging land, trade, economic, and development policies. The former is deemed to be legitimate action within the field of biology, while the latter is deemed as politics and outside the fold. Yet as we have seen, biology and politics are forever embrangled through colonial afterlives. Environmentalists fight for a nostalgic past even while destructive economic land policies continue to shape the landscape and the environment. Land policies have been disastrous for native infrastructures, as America's poor neighborhoods, indigenous nations, and communities of color have long known. Similarly, global neocolonial policies of the World Bank and the International Monetary Fund have meant that poor nations have been stuck in cycles of debt-ridden economics even after independence.[39]

If colonial nostalgia and settler logics haunt invasion biology, might the advent of climate change upset some of these settled wisdoms? After all, as native environments are transformed by climate change, what does it mean to be native anymore? The idea of good native and bad foreigner has been critiqued within biology.[40] Moreover, native plants can also become invasive,[41] and foreign species can develop new ecological networks of botanical and political geographies. I am repeatedly amused by the response of biologists. For example, some have invented a new term for the migrating response of species, *neonatives*, a new category to describe species that have moved at least hundred kilometers or a few hundred meters beyond their native ranges established in 1950.[42] It appears that there is a strong impulse to hold on to the idea of the native.

Invasive species rhetoric is grounded on fixing plant biology into a moment of the past, as though species stop adapting and evolving, creating new relations and relationalities. In fact, plants and animals migrate, leading to astonishing new adaptations and even accelerated evolution; once-foreign plants can integrate, assimilate, help, and even become indispensable to "native" communities.[43]

What Is Native?

The idea of "nativeness" is a recent invention. Its definition lacks "fundamental biological meaning" and remains "theoretically incoherent."[44] Despite a lack of a clear, coherent, or consistent definition of this central term, invasion biology soldiers on. Broadly, biologists recognize that species move and that migrations of plants are central to plant biology and theories of plant ecology and evolution. However, once humans become agents of that move, then we see the emergence of a new set of proliferating terms—*exotic, nonnative, introduced, alien, naturalized*.[45] While not always consistent, much terminology in Europe and the Americas revolves around 1492, when life forms from the Americas were brought in earnest to European shores. Yet many Asian and African imports arrived in Europe before Columbus visited the Americas—for example, cypress trees were brought from the Middle East to Italy in classical times, and rats followed Romans across Europe. Often the term *naturalized* is used for species not originating in Europe but living there before 1492.[46] In Australia, the focus is 1770 or 1788, years that mark the arrival of British colonizers into Australia.[47] But exceptions are abundant.

In the context of the long arc of colonialism, I want to address three recent arguments attempting to resurrect *invasive species* as a progressive term.

First, within plant ecology, new entrants into an ecosystem are dubbed *colonizers*. In such a definition all migrants, refugees, and nomads are colonizers. Recently, some want to extend the term to invasive species as well, arguing that distinguishing between human-mediated biological invasions and those that involve more natural colonization of plants is a false binary.[48] After all, many invasive species were brought by humans decades ago but have only turned invasive recently. Whereas *invasive* evokes xenophobia, *colonizer* perhaps better captures the aggressiveness of the plant. Within such a framing, the term *invasive* thus places blame on the plant. Reframing invasive species as colonizers, in this view, is a project of anticolonialism.

Others argue that to be against invasive species is to engage in botanical decolonization. For example, Mastanak and colleagues rethink native plants as a discursive field where vocabularies of native, nonnative, and nativism can be interrogated around a politics of "botanical cosmopolitanism."[49] Using this analysis to think through colonial links, they argue that invasive species can become a site for a "broad process of botanical decolonization and a strategic location for ethical action in the Anthropocene."[50] To their credit, in their analysis they go back to the early stages of European territorial expansion in the sixteenth and seventeenth centuries, noting that modern gardens and lawns are also part of the colonial project. Arguing that we should understand colonialism as a multispecies project that involves the settlement of plants, they disavow Edenic narratives. To them, home "native" gardens are an explicit challenge to the primacy of "lawns" and an act of domestic conservation.[51]

Yet others reframe invasion biology as local extermination. For example, Keulartz and Weele extensively discuss different framings of invasion biology.[52] Instead of deploying xenophobic, racist, and belligerent images, they see the project as protecting the local.[53] Like global superstores such as McDonalds and Walmarts that destroy local businesses, so too do foreign species outcompete and destroy local ecologies and local cultures.[54] In short, rather than focus on invaders as the inherent problem, we should focus on their impact on local cultures through the homogenizing forces of globalization. Foreign species threaten the local; we should protect local species diversity because of their value. We shift the concern from the foreign and invasive species per se to safeguarding the local.

Rather than reframe invasion, all three arguments—invasive species as colonizers, invasion biology as decolonizing, and the preservation of the local native—reproduce some of the key politics that animate invasion biology. They hinge on a tension between indigenous studies and diaspora/postcolonial studies. Are invasive species a legacy of migration and hybridity (in diaspora and postcolonial studies) or a figure of settler colonialism, the violent colonizers of the native (indigenous studies)?

Who is the "native"? To me, the "native" within the concept of invasive species is a native of settler colonialism and not the "native" of indigenous studies. Let us consider each of the three arguments.

First is the idea of invasive species as evil colonizers. In transferring meanings across human and plant words, something significant is purposefully misunderstood and elided. Just as the vocabulary of natives and aliens ignores the histories of colonialism that reshuffled the world's biota, so too does the term *colonizer* misrepresent a history of invited invasions. While the ecological sciences incorporated the term *colonization* as an action of a plant or animal that establishes itself in a new area, within histories of colonialism its meaning is entirely different. The heart of European colonization is not just people moving into new areas and establishing themselves but fundamentally an extractive economy—they looted the colonies' rich resources to aggrandize the metropole. As Walter Rodney argues in the case of Africa, colonists "developed" Europe by "underdeveloping" Africa. This can be said of most colonial contexts. In the case of India, colonialism decimated one of the richest and most industrialized economies of the world (which together with China accounted for almost 75 percent of world industrial output in 1750) into one of the poorest, most illiterate, and most diseased societies by the time of independence in 1947.[55] Colonialism not only exported resources but also exploited people through slavery and indentured servitude. To foist this vast legacy of colonialism onto the plant world is simply perverse.

The second argument is about land. As Leanne Simpson outlines in one of the epigraphs to this chapter, land is not an enclosure protected by a border, a natural resource, or a commodity.[56] Land in indigenous studies is a space of affective ecologies, relationality, and attachments nurtured over long periods of time. In contrast, the "native" of invasion biology is the "native" of settler colonialism, of stolen land, commodification of plants, and short-term profits; this is the space that botany inhabits. Until we seriously contend with questions of land, stop relating to plants as research subjects and objects of study or property, and stop appropriating and exploiting vegetal worlds, botany is a colonial project. Until we understand plants as coinhabitants of planetary life and build affective relations with plant worlds, appropriating the language of indigenous studies only reinforces the settler colonial project.

Finally, we come to the question of the local. Shifting the focus to protecting local culture only reproduces and exacerbates the more troubling aspects of an already reactionary framework of bad foreigner/good native.[57] The anti-imperial approach corroborates rather than contradicts the rigid dichotomy between the purity of local "authentic" cultures (i.e., botanical nativism) and the corrupting and contaminating influence of outside forces (i.e., botanical cosmopolitanism). The underlying assumption that globalization is homogenization has been challenged in postcolonial studies. After all, globalization also offers sites of refuge and hybridization, mixing and blending of identities that lead to new forms of diversity in human and nonhuman worlds. Nativism and cosmopolitanism are false dichotomies for human and vegetal worlds alike. Within invasion biology, nativeness is revocable while the nonnative is permanent.[58] I remain unconvinced by the various proposals that work to resurrect invasion biology by other names or frames. It remains a singularly white settler colonial project.

Cultivating Living Relations

The constant traffic of flora and fauna across nations through agriculture, animal husbandry, forestry, and other national and international commercial ventures have created complex local and global ecologies of relationality. Indeed, a whole burgeoning subfield of multispecies ecologies, closely related to indigenous, feminist, and queer ecologies, has proved fertile ground to explore the complex interrelations that have developed through colonial, settler-colonial, postcolonial, and neocolonial times. It is not a simple story but one that is infinitely complex, embrangled, and fascinating.

My conclusion from this work is that the "native" of indigenous studies is fundamentality about relationality, about and among living relations. In contrast, the "native" of invasive species and settler colonialism is one of life in enclosures, a relation of alienation. Nowhere in the writings on invasive species is there any fondness of foreign species, any affective connections, even as humans rely on alien species for most of their food and basic requirements.[59] Kim TallBear reminds us not to fetishize or fossilize the indigenous.[60] The "indigenous" is never static but always changing and evolving, dynamic like everything else. Cattelino draws parallels between cultures and natures and notes that both can emerge as static categories of a bygone era, relegated to a mythical and pure past.[61] My point here is not to create analogies between peoples and plants but to understand how both plants and people are circumscribed into "colonial timelines."[62] We need to understand that settler colonialism and foreign species are both being framed in the field of invasion biology by and through settler colonial logics.

Much recent work on indigenous ecologies makes this point. As Jessica Hernandez argues, indigenous sciences have a different view of invasive species.[63] Western sciences label invasive plants as pests, unwanted, or not belonging to the landscape. Yet, most invasive species were in fact introduced during colonial times by settlers and colonizers. White settlers have assimilated in the Americas and lost connections with their plant kin, and they now deem those very species as terrible beings.[64] Botanical amnesia surfaces again. As Hernandez argues, there is no word for conservation in many Native and Indigenous languages because there is no need. One does not exploit the land, destroy it, and then look to conserve or restore it! The plant sciences must begin with ethical relationships with the land and its many inhabitants.

Speaking directly to this point, Kim TallBear argues that American dreaming, even in its inclusive and multicultural tones, nonetheless is grounded in the elimination of Indigenous people.[65]

> Thinking in terms of being in relation, I propose an explicitly spatial narrative of *caretaking relations*—both human and other-than-human—as an alternative to the temporally progressive settler-colonial *American Dreaming* that is ever co-constituted with deadly hierarchies of life. A relational web as spatial metaphor requires us to pay attention to our relations and obligations here and now. It is a narrative that can help us resist those dreams of progress toward a never-arriving future of tolerance and good that paradoxically requires ongoing genocidal and anti-Black violence, as well as violence toward many de-animated bodies. . . . A relational web framework can also articulate obligations across the generations, or over *time* if one is attached to that idea.

Reo and Ogden highlight the impoverished vocabularies of borders, property, and ownership that undergird settler science.[66] In contrast, the Anishnaabe people regard plants, like all beings, as persons that assemble into nations more so than species. Rather than pathologizing them as invasive, the Anishnaabe teachings regard new plants and animals as part of natural processes resulting from migration by other-than-human nations. Plant movement is not inherently good or bad. According to Anishnaabe teachings, it is the responsibility of humans to determine why the migration has occurred, sometimes with the assistance of animal teachers. Linking environmental change directly to the introduction of a Euro-American land ethic, Anishnaabe teachings see culpability in invasive ideologies rather than in specific animals or plants.

Together these framings allow us to see how much power we cede to the temporal horizons of European colonization as biologists.[67] We can address the colonial histories of botany by discarding settler logics and developing different relations to the land by recognizing its stolen histories. Only in refusing our botanical amnesia and rememorying the hauntings of colonial biogeography might we build more just worlds.

CHAPTER EIGHT

Like a Tumbleweed in Eden

Diasporic Lives of Empire

Until the lions have their own historians,
the history of the hunt will always glorify the hunter.
CHINUA ACHEBE | interview in the *Paris Review*

No fiction, no myths, no lies, no tangled webs—this is how Irie imagined her homeland. Because *homeland* is one of the magical fantasy words like *unicorn* and *soul* and *infinity* that have now passed into language.
ZADIE SMITH | *White Teeth*

THE CAMERA OPENS with an eerie, dusty, desolate street bathed in the shimmering hues of a dusky evening. Rows of squat wooden buildings frame each side. The wind howls. The music plays discordant notes of foreboding. Then the quintessential icon of the American West makes its appearance: a large brown mass of dried, entangled plant matter rattles down the street. The camera follows the rattling ball as it rolls and bounces on and on, seemingly endlessly with strange precision, a purposeful gait, and an enlivened spirit. This scene is the classic archetype of the American West, and the hero, the rattling ball, is the floral icon, the classic tumbleweed. The structure houses thousands of seeds that scatter about as the tumbleweed mass bounces along its path. It is worth noting that the tumbleweed, an icon of the American West, is not native to the United States or even the Americas. Indeed, none of the classically quintessential American icons—the cowboy and his accessories of horse, cattle, and tumbleweed—are native to the United States.[1]

As Zadie Smith suggests in one of the epigraphs to this chapter, ideas of homelands are mythic, imagined concepts, purposefully created for imagined pasts, presents, and futures.[2] The tumbleweed, in fact, is Russian in origin, commonly called Russian thistle, brought to South Dakota by Russian religious refugees in the 1870s. And ever since, "its phantom path" has grown "green with offspring" as it tumbled its way into the rest of the country.[3] Why and how do some objects come to be beloved and to inhabit the iconography of a nation—tumbleweeds of the West, the cherry blossoms of Washington, DC, the Georgia peach—while other objects remain resolutely alien, foreign, and reviled, such as kudzu, the Asian long-horned beetle, and zebra mussels?

In this chapter, I explore concepts around mobility and "migration" and how the values and political contexts accompanying these concepts circulate across geopolitical and botanical terrains. Within such interdisciplinary frameworks, surprising stories emerge of where the concepts first originated and how they traveled to new contexts, only to return to reanimate the original contexts with new cultural, political, and biological valences. Questions of land and belonging are particularly difficult for migrants. Three generations ago, my ancestors moved because of a famine. I moved for a love of science. Politics of belonging are uneven, unequal, and difficult in histories of colonialism. In further extending theories of migration to examining the coevolved histories of humans and nonhumans, nature and culture, we witness the embrangled histories of disciplines and their concepts and the interwoven histories of multispecies lives and travels.

After all, as with the quintessential tumbleweed, this question of what comes to belong and what does not is much more complex than mere geography.[4] While the topographies of national belonging might be varied and complex, many plants and animals, like humans, have a global presence, having been scattered to the ends of the earth through the work of diaspores and diasporas. In this chapter, I focus on the twin conceptual power and utility of the concepts of diaspore and diaspora to highlight the shared and embrangled roots/routes of our natural and cultural worlds, or naturecultural worlds.

Diaspores and Diasporas

Diaspora, n: The movement, migration, or scattering of a people away from an established or ancestral homeland.

Diaspore, n: In botany, the dispersal unit consisting of a seed or spore plus any additional tissues that assist dispersal.

Diaspora is an ancient word that has since been extended to other peoples and contexts.[5] *Diaspore* and *diaspora* both come from the Greek verb *diaspeiro*, "to scatter about, or to disperse around." While the words are related etymologically, here I explore their conceptual and interdisciplinary interconnections. *Sper* implies "to scatter" and is used to describe both the scattering of seed and the dispersal of people. In theorizing the worlds of movement, mobility, and migration, I find the concept useful for several reasons. It is useful across human and nonhuman worlds and helps us understand that nature and culture are not binary opposites but rather worlds with intertwined histories linked not only historically and conceptually but also etymologically. The word is increasingly connected to the ravages of colonialism and consequences of globalization—two processes central to understanding the travels of human, nonhuman, and conceptual worlds. It links the geographies of the colonial and colonized worlds, and scales of the global and local.

Unlike such terms as *rhizomes* or *rhizomatic networks*, which refer to networks of clonal plants that proliferate only through asexual means, *diasporas* invokes sexual and asexual propagation through processes of recombination, adaptation, and mutations.[6] *Diaspora* is by definition always relational, linking migrating communities and forever binding them to the lands and ancestors they left behind, but never collapsing the two into some essentialized mythical past. While terms like *nomadic* and *traveling concepts* also imply mobility, diasporas incorporate the notion that branches of migrating groups also dig new roots in the land of their immigration, forging new ecologies and economies.[7] Finally, while dispersing plants and populations are related to the original population, diasporas allow for more imaginative sexualities—the possibilities of admixture and cross-fertilization through a politics of assimilation, or of remaining resolutely unique and distinct through a politics of isolation and separatism. Diasporas open up robust and imaginative landscapes to trace the embrangled conceptual histories of multispecies lives and travels.[8]

Sensing Like an Empire

Globalization is the imposition of the same system of exchange everywhere. In the gridwork of electronic capital, we achieve that abstract ball covered in latitudes and longitudes, cut by virtual lines. . . . The globe is on our computers. The planet is the species of alterity, belonging to another system.
GAYATRI CHAKRAVORTY SPIVAK | *Death of a Discipline*

I draw on "sensing" as a metaphor to examine the central role colonialism has played in shaping the biotic landscapes of the world.[9] As Chinua Achebe astutely observes in one of this chapter's epigraphs, those with power, a voice, and access to the historical records are ultimately the ones who write history.[10] It is remarkable how colonialism's role in reshuffling the world's biota has been obscured in our current environmental anxieties about invasive species. Thinking about colonialism and botany through the lens of diaspora is illustrative. In previous work, I have explored the field of invasive species management, arguing that we cannot understand how the field and its allied practices and policies of conservation and restoration ecology have developed without recognizing the inextricable interconnections between nature and culture.[11] In our quest to be modern, and in our claims of modernity, we have vivisected a vibrant organic planet and the dynamic knowledge systems it has generated into narrow, parochial, and insular disciplines.[12] What has become obscured in our disciplinary silos is that plants, peoples, and ideas are inextricably entangled. To understand the coconstitution of these worlds and the coproduction of knowledge, we need to examine them together.

Over the past three decades, feminist and postcolonial critics of science have elaborated the relationship between our conceptions of nature and their changing political, economic, and cultural contexts. Naturecultures are important because they reveal the centrality of the history of colonialism and the enduring legacy of the colonial imaginary on our knowledge systems, helping us think about diasporic entanglements across species and across our knowledge systems.

Indeed, as species move, so do theories, ideas, and concepts. Concepts migrate across disciplines—from the sciences to the humanities and back—and are repurposed to theorize new objects in new contexts. Many terms span species and disciplines—from human contexts in ethnic studies and postcolonial studies to scientific/biological terminology: *alien*, *diaspora*, *native*, *local*, *foreign*, *colonizer*, *colonized*, *naturalized*, *pioneer*, *refugee*, *founder*, *resident*, and *exotic*, to name a few. Indeed, environmental historians have long argued that rather than thinking of the impact that colonialism has had on biology, ecology, or the environment, it would be more accurate to think of imperialism as fundamentally an ecological project, in which humans, plants, and other species were shuffled around the earth in schemes for colonization and conservation.[13] Colonialism was the original bioinvasion of mammoth proportions. In the early modern period, botany, for example, was big science and big business, critical to Europe's ambitions as a colonial trader. Understanding the field of botany through colonial historiography reminds us it was never the pure story of rigorous and apolitical taxonomies, nomenclature, and systems of classification but rather a science deeply embrangled in colonial ambitions.[14] Alongside this movement emerged fundamental concepts in ecology, evolution, and other biological and human sciences.

We have historically imagined our relationship with the biota of the world in numerous and diverse ways. In his influential book *Ecological Imperialism*, Alfred Crosby argues that the roots of Europeans' domination of the Western world are in their creating "neo-Europes" wherever they went, especially in North and South America, Australia, and New Zealand, where settler colonialism ravaged native populations of humans, plants, and animals. Rather than thinking of European domination as the result of technology, Crosby argues that we should understand it as a simultaneously biological and ecological project. The colonists took with them a "portmanteau" of biota—plants, animals, and pathogens—that enabled expansions of Europe and a radical transformation of the globe. Where Europeans went, their agriculture and animals went; they thrived while indigenous ecosystems collapsed. This vast migration of species ushered in a bioinvasion of mass proportions by the conquerors' animals, plants, weeds, and germs, yielding a "great reshuffling."[15] Some plants were now ubiquitous not just in Europe but around the globe. As Crosby remarks, "the sun never sets on the empire of the dandelion."[16] The science of breeding and horticulture led to scientific breeding stations scattered around the globe that turned raw materials from the colonies into plantation crops for the British, French, and other colonies and empires, including the Americas.[17]

Indeed, it is more accurate to talk about our entire planet as having been biotically reconfigured due to this long history of what Richard Grove calls "green imperialism."[18] Colonial expansion ushered in not only an unprecedented movement of people, plants, and animals but also concerns about ecological destruction and degradation. If ecological exploitation has its roots in colonial expansion, Grove argues, so does environmentalism. Conservation biology emerged as a form of green imperialism, with a vision of Edenic islands that were being harmed through colonial extraction. The tumbleweed was now in Eden. The diasporic scattering of tumbleweed seeds and the vision of Edenic ecologies are both central to utopic ecological thinking. This mindset allows us to understand the etymological definition of *diaspora*—of how Europeans helped disperse and scatter plants, animals, and pathogens and thus enabled the expansion of Europe. At the same time, in their desire to recreate "homes" or little "Europes" in the colonies, they transported European biota to restructure the colonized landscape, thus destabilizing our association of flora and fauna with the natural or "autochthonous landscape."[19] A recent study confirms that the redistribution of species accelerated with the start of European colonialism and that the colonial impact is still detectable in alien floras worldwide.[20] They find, for example, that the composition of plant species is more similar across colonies occupied by the same empire. In an ultimate act of irony, they now insist on preserving the "neo-Europes" from newer immigrants from Asia, South America, and Africa.

The free and profligate movement of biota during colonialism reminds us of the hubris of empire, where colonists carried their landscapes with them, building new landscapes in their new colonized lands and showcasing colonial flora and fauna as exotic items in their homes and museums.[21] An open and laissez-faire policy toward plant and animal worlds shifted, in part, because of changing relationships with nature. In the decades after the US Civil War, industrialization, urbanization, and westward expansion transformed the nation's landscapes and redefined Americans' relationship with nature.[22] The emerging love of nature was evidenced in the dramatic growth in the number of Americans who considered themselves "nature lovers," and Americans saw their love of nature as the quality that distinguished the "natives" from newer immigrants. A love for nature translated into a zeal to protect it, and (newer) immigrants came to be seen as not loving nature and therefore as a problem. Nativists increasingly challenged federal government passivity about immigration. The "native" emerges as the sole site of "purity" in our conceptions of humans and plant and animal ecologies.

The Language of Empire

Natural and cultural worlds are always mediated not only through the material bodies of living organisms but also through discursive practices and language.[23] Theories of natural history and much of botanical nomenclature and organization are best understood through colonial histories. Indeed, conservation biology also has its roots in colonial thought. Since so much of colonialism was about resource extraction, colonies' material resources were vital to empire. Indeed, some of the earliest conservation projects were outside of Europe and in the colonies, produced by a coterie of professional scientists who worked to prevent the depletion and hasten the renewal of the resources in the colonies.[24]

The colonial legacy of botany is immense and deep. Let us start with the very nomenclature. As Janet Browne argues: "Just as the British Museum and Kew Gardens were constituted by the flora, fauna, and human knowledges extracted from the colonies, the discourse of natural history was articulated in terms of biotic nations, kingdoms, and colonists, reflecting the 'language of expansionist power.'"[25]

Classification and nomenclatures are deeply rooted in the politics of their times. In the wonderful essay "Why Mammals Are Called Mammals," Londa Schiebinger reveals that when Linnaeus was naming his classification system, there was a big campaign in the United Kingdom to promote breastfeeding.[26] Therefore, even though mammals are defined by many characteristics that distinguish them from those assigned to other classes, such as hair, a four-chambered heart, a single-boned lower jaw, three middle ear bones, a diaphragm, and mammary glands, and although all mammals maintain a constant body temperature, it is the feeding of the young that came to define us. It is hardly accidental that in a cultural context where nature is female, the focus on breastfeeding bolsters cultural notions of self-definition as a species where women are the caregivers. Such cultural ideas and hierarchies pervade taxonomies of flora and fauna. The idea of a "great chain of being" explicitly organized species as a hierarchy through an episteme of difference—humans on top and lowly plants at the bottom. More particularly, in this episteme, the white man (more precisely, the white heterosexual couple that included the white man's complement, the white woman) was always at the pinnacle of the ladder.[27] The colonial nomenclature of life folded in fundamentally determinist discourses of race, gender, and nature. As colonial expansion permeated the globe, Western systems of nomenclature supplanted local knowledge. Colonial legacies have dismantled and erased a plethora of languages, meaning-making practices, and nomenclatures to usurp multiple cultures of knowing, replacing all with a universal, scientific "monoculture of knowledge."[28] Such erasure is violence, and "this legacy of capturing and renaming nature leaves the postcolonial writer in the position of having to renegotiate the terms of taxonomy, struggling to articulate new relationships and new meanings in the tired language of empire."[29]

The intricate web of migrations of plants, animals, and humans—of ecological migrations through European colonialism and ethnic diasporas—have led to the fields of environmental humanities and postcolonial ecology, both of which foreground the historical process of nature's mobility, transplantation, and consumption. As Elizabeth DeLoughrey and George Handley argue, an ecological frame is important because (1) geographies have been altered by colonialism; (2) environmental dualisms we recognize and know so well, such as culture/nature or male/female, were in fact constituted through the colonial process; (3) environmentalism also has colonial roots, and postcolonial criticism has effectively renewed, rather than belatedly discovered, its commitment to the environment; (4) in human/nonhuman binary in Western thought, nonhuman worlds have always been conceived as the binary Others of Western man.[30] To understand the phenomenon of anthropocentrism, we need to engage with nonhuman worlds and deep time, always remembering that the *anthropos* is inevitably Western man. While botany's development and its nomenclature, theories, and practices were constituted through the expansion of empire, the conceptual landscapes of botany traveled to other sites. Nowhere are these colonial roots so evident as in tracing the diasporic lives of concepts. Through an example of invasion biology, I briefly track the evolution of the recent history of the native/alien binary and explore how the geographic binary is best understood alongside the twin conceptual terms of *diaspores* and *diasporas*. I highlight the complex circulations of biota and their attendant vocabularies while noting some patterns in the travels of ideas, theories, and concepts.

THE POLITICAL UTILITY OF CONCEPTS

Diaspores and diasporas are caught in the potent politics of contemporary immigration. Let us begin with the example of invasion biology. Here I draw on my previous work on invasion biology.[31] The idea of invasion is predicated on a discourse of "nature in place" and "nature out of place"—and by definition, invasive species are nonnative species that are introduced where they do not belong. While "native" species may get out of control and begin to dominate the landscape, they, by definition, will never be "invasive." This idea of "nature in place" has a complex and nonlinear history. The concept of "nativeness," first introduced by the English botanist John Henslow in 1835, was subsequently used by Hewett Watson to delineate "a true British flora." Watson's terminology drew on English common law about human citizenship rights.[32] Of course, determining the "true" British flora simultaneously defines the "not true" British flora, and thus arose the now-familiar binary of the native/alien. While the terms continued to be used in the coming decades, no general policy about natives/aliens emerged.[33] It is only with growing nativism that we see the beginning of the policing of borders to all living organisms entering and leaving nations. In the United States, as Philip Pauly notes, the paradigm of the nativist approach was the Chinese Exclusion Act passed at the insistence of California workingmen in 1882, a year after the state's quarantine law.[34] After World War I, Congress introduced limitations on entries of all European immigrant groups through the Immigration Act of 1924. Again, we see that human, plant, and animal histories are inter twined. Indeed, ideas of nativeness and the closing of borders emerge first with respect to plants and animals before being given legal and institutional life in the management of people through the state and its borders.

THE NATURECULTURAL LIVES OF CONCEPTS

As the example of the origin of the native/alien binary shows, concepts, once they emerge in one sphere, are often taken up in other spheres, so the two grow inextricably interconnected. Modernity presupposes a distinction between nature and culture and between human and nonhuman even while history demonstrates that these are false binaries.[35] My work in tracking the language of *native* and *alien* demonstrates how xenophobia pervades both terms. Here I present a brief summary of my previous analysis of how invasive species are represented in the media.[36] Let us briefly consider some of the language in the United States around foreign/exotic species. Consider some of the headlines from newspapers and magazines:

- Alien invasion: They're green, they're mean, and they may be taking over a park or preserve near you
- Aliens wreaking havoc
- Native species invaded
- It's a cancer
- Creepy strangler climbs Oregon's least-wanted list
- Biological invaders threaten U.S. ecology
- U.S. can't handle today's tide of immigrants
- Alien threat
- Biological invaders sweep in
- Stemming the tide of invading species
- Experts warn of the growing invasion of foreigners into the nation's aquatic systems
- Invasive species: Pathogens of globalization

If you read them carefully, most of these headlines do not specify that the accompanying article is about plants and animals; rather, the titles present a more generalized classic fear of the outsider, the alien that is here to take over the country—a vision of how we are moving from peaceful, coevolved nature to an uncertain future with alien and exotic plants and animals.

As I examined the conceptual universe of our discourse around foreign plants/animals and foreign peoples, the similarities were striking. Like the germ panics surrounding immigration and immigrants, questions of hygiene and disease haunt exotic plants and animals.[37] Like the "unhygienic" immigrant people, alien plants are accused of "crowd[ing] out native plants and animals, spread[ing] disease, damag[ing] crops, and threaten[ing] drinking water supplies."[38] The xenophobic rhetoric that surrounds immigrant humans is extended to plants and animals.

We can see distinct parallels in this rhetoric between foreign humans and plants and animals. First, aliens are other. One *Wall Street Journal* article quotes a biologist's first encounter with an Asian eel: "The minute I saw it, I knew it wasn't from here."[39] Second, alien/exotic plants are everywhere, taking over everything—an unstoppable march from national parks to our backyards. Third, there is a suggestion that they are silently growing in strength and number. If you haven't heard about biological invasions, it is because "invasion of alien plants into natural areas has been stealthy and silent, and thus largely ignored."[40]

Fourth, aliens are difficult to destroy or eliminate and will persist because they can withstand extreme situations and are indestructible. Fifth, they are seen as "aggressive predators and pests and are prolific in nature, reproducing rapidly."[41] Alien plants are repeatedly characterized as aggressive, uncontrollable, prolific, invasive, and expanding. As one article summarized it: "They Came, They Bred, They Conquered."[42] The oversexed female is one of the classic metaphors surrounding immigrants. Foreign women are associated with super-fertility, reproduction gone amok. Along with the super-fertility of exotic/alien plants is the fear of cross-fertilization. Native species in this story are passive, helpless victims of the sexual proclivity of foreign/exotic males. Sixth, once these plants gain a foothold, they never look back. Singularly motivated to come and take over native land, they remain unconnected to their homelands; it is implied that they will never return and are therefore "here to stay." Like portrayals of human immigrants, the greatest focus is on their economic costs because they are believed to consume resources while returning nothing. Finally, feeding on the images of illegal immigrants arriving in the country through difficult, stealthy, and arduous journeys, exotic plants and animals are accorded the same metaphors of illegal, unwelcome, and unlawful entry.

THE POLITICS OF NAMING

Indeed, the pervasive xenophobia felt for humans, plants, and animals has not gone unnoticed. Since September 11, 2001, fierce critics have challenged the growing xenophobia in the United States. Ideas from immigration activism drawn from analyses in the social sciences and humanities have been embraced by a wide group of biologists and in well-publicized locations. Hugh Raffles wrote an op-ed in the *New York Times* titled "Mother Nature's Melting Pot," and then Mark Davis and nineteen other biologists, in a commentary in the journal *Nature*, cautioned, "Don't judge species on their origins," challenging the xenophobic underpinnings of invasion biology.[43] The biologists contended that "'non-native' species have been unfairly vilified for driving 'beloved native' species to extinction, thereby creating 'a pervasive bias'" against alien species.[44] They argued that a dichotomy between native and alien was becoming less useful and even counterproductive and recommended that we abandon such thinking.

These articles make several important points to remind us that focusing on alien plants as a problem obscures many facts. The natural world is fast changing because of many unrelated factors such as climate change, land use changes, overdevelopment, and ecological degradation. Most campaigns to eradicate invasive species have simply not worked. And new arrivals can even help an ecosystem rather than hurt it or can have a mixed impact. For example, in California, some native butterflies feed on nonnative plants. In Puerto Rico, alien trees have restored abandoned pastures, making them suitable for native plants.[45] Agriculture in North America would be very different if only native plants could be grown without the services of the European honeybee as pollinator.[46] Thus, the anti–invasive species campaigns mischaracterize the native/alien distinction. Most Americans do not realize that many of their prized flora and fauna are foreign in origin. For example, the state birds and flowers of several US states migrated from other parts of the world. In contrast, some native species have proved to be invasive and have caused great damage. The categories of native and exotic house too much diversity to be useful anymore.

Classifying organisms by their "adherence to cultural standards of belonging, citizenship, fair play and morality does not advance our understanding of ecology,"[47] note Davis and colleagues. Instead, we ought to embrace a more dynamic and pragmatic approach, focusing on the function of species in their ecosystem rather than a litmus test on their geography of origin. Plants and animals, like humans, need a "thoughtful and inclusive response."[48] Yet in a post-9/11 United States, Homeland Security and invasive species activists do very similar work in calling for the naming of the foreign as a threat. As human anti-immigration activists have increasingly called for identifying and deporting human immigrants, many environmentalist groups have also called for identifying and exterminating foreign and invasive species. Both sides agree that only a fraction of the "foreign" species are harmful, but all agree those that are harmful can be very destructive and need to be reined in. But the crux of the issue is, how do you tell which alien will become "invasive"? Nip it in the bud, some biologists say—catch it before it becomes a problem. Similarly, how do you tell which alien will become the terrorist? Again, immigration activists suggest nipping it in the bud— proactively targeting particular nations and racially profiling individuals, whether they are citizens or not.

Diaspores and diasporas are caught up in analogous potent politics of xenophobia in the "war on terror" and the notion of "invasion biology." Fundamentally, these debates are about the politics of nativism, national aspirations, and the imaginations of who belongs and who does not. Understanding and tracing these disparate arguments and concepts in biology and politics through a naturecultural lens allows us to see the common conceptual terrain that diaspora/diaspore fields share through centuries of coproduction and cross-pollination. Diasporas and diaspores present us with a powerful and capacious concept that enables interdisciplinary and intersectional naturecultural histories. It reveals the impoverished accounts of diaspores and diasporas that separate human and nonhuman worlds. It is the concept that reminds us of the hubris of empire and the irony of contemporary immigration politics. It gives us a way forward for our lives in the ruins.

Embracing Contradictions Within: Tumbling Together

So bring back the days of my emerald wanderings
Dream in a wide open sky . . .
Roll me like a tumbleweed in Eden
All the way home
CHRIS ROBINSON | "Like a Tumbleweed in Eden"

"Like a Tumbleweed in Eden" is the evocative title of a song by Chris Robinson of the Black Crowes. It presents an anachronistic vision of a dried, weedy rattle scattering its seeds in a pristine mythical Eden. The contradictions are everywhere. A foreign plant is an American icon. Edenic visions are at once mythical and all too real. Our rigorous, rational, and supposedly apolitical theories of nature are bound in binaries of nature and culture and deeply entrenched in histories of colonialism; indeed, they are constitutive of colonialism and their attendant politics of race, class, gender, and sexuality. Nature is forever represented as "female" and, at least in the Western colonial imagination, an entity to exploit. Indeed, reading Western imaginations of the environment "is like watching a spectacular dramatization of heterosexual teleology."[49]

The conceptual tools of naturecultures and diasporas allow us to unravel the profound mobility and migration that colonialism has wrought. As a diasporic biologist now teaching in a college of humanities, a diasporic South Asian now living in the United States, diasporas are for me also personal. In the opening lines of the haunting poem "To the Diaspora," Gwendolyn Brooks writes[50]:

> You did not know you were Afrika
> When you set out for Afrika
> you did not know you were going.
> Because
> you did not know you were Afrika.
> You did not know the Black continent
> that had to be reached
> was you.

Brooks argues that in her search for Afrika, she discovers that what she was seeking was in fact always right there in front of her—in fact, within herself. I find this deeply resonant. Trained as a biologist, I turned to the humanities to understand the politics of difference. As it turns out, the politics of difference was always deep within biology—in the very materiality of living organisms and in the vocabularies and concepts of the field. In understanding the links of diaspores and diasporas, I have come to discover the potent worlds of naturecultures. Indeed, the tumbleweed belongs in Eden. Only, rather than the mythical place of the Western imagination, Eden is instead a place teeming with diversity and difference, attuned to the ravages of colonialism and slavery and calibrated to histories in deep time. The tumbleweed in Eden rattles on and along the way scatters seeds for less-ravaged futures.

INTERLUDE

Fables for the Mis-Anthropocene: Love the Dandelion

The wealthy suburb of Waisley was a liberal town, high on the whiffs of environmentalism. Of cities in the United States, they had the highest numbers of solar panels, renewable energy generation, composting, and energy-efficient buildings, and the lowest per capita water consumption. They had majestic, community-supported agriculture gardens and fruit orchards. A wealthy district with high property taxes, it also produced the country's best schools. When the Raucous Spring erupted around the world, children raised with environmentalist zeal took up the charge.

Chirp-Net had recently been shaken. Indigenous children the world over wrote stories about stolen land and the horrors of settler colonialism. These stories disabused any notion of colonialism as a thing of the past. Children confined to reservations and refugee camps wrote about the inhumane conditions they lived under. The Chirp-Net gave voice to young, vibrant, inventive children who had spent most of their lives in exile. The contrast was unbearable for other children. Each day, a steady stream of stories chronicled extinctions across the world. Members of the Chirp-Net leapt into action. While they worked on the difficult issues of nations and nationalism, they took to heart that all radical actions begin at home—they attended to the xenophobia in their own homes and gardens.

Waisley, like many towns across the country, was in the throes of panic around invasive species. The town had outlawed the planting of certain foreign plant species. So worried was the town that it enacted laws stipulating that at least 80 percent of plants in home gardens be native species. Nationalism in the front yard! As it turned out, Jessie of Michigan posted a superb manifesto about invasive species on the Chirp-Net. Why were some plants and animals declared foreigners, exotics, and invasives? Jessie had witnessed the uproar around the invasive Asian carp in the Great Lakes, the same species that was overfished and endangered in China, where they were a delicacy. After all, Jessie wondered, what is a native species? Why did the white settler colonists not consider themselves invasive too? What exactly did native mean, then, if not a convenient political invention to keep power? They described how nativism in Nazi Germany had brought fascism to their gardens by regulating what could be planted in them.

Humans had moved plants and animals across the globe during colonialism. Why blame plants? We are expected to make our communities welcoming to newer people, Jessie argued. Why do we not extend this to all creatures on earth? After all, with rapid climate change, creatures are no longer adapted to their birth homes. What does "native" mean in the age of climate change? Humans can turn on the air conditioner or the heater, but plants and animals cannot—they have to move, to migrate in order to find more suitable environments for themselves and future generations. Jessie expressed outrage that western colonial nations once colonized the world with impunity, wildly moving plants, animals, and humans around, but were now refusing to acknowledge this history. Instead, western nations defeated by colonial nations fortified the homeland by building more and more borders, boundaries, walls, and laws to prevent the entry of newer humans, plants, and animals. Hypocrisy, Jessie cried.

The "good life" of suburbia relies on various histories of domination of nature. The children of Waisley took this up. They resurrected an old movement to do away with lawns. Homes in Waisley, like many wealthy towns, sported expansive lawns. In England, where this practice came from, lawns were the manifestation of a wealthy estate. The United States had embraced this status symbol and democratized it. Well-kept lawns were everywhere, a sign of wealth, prestige, and privilege. Gardening clubs and horticultural societies ran popular yearly contests. It was an honor to be chosen, and competition was strong. But the Abolish the Lawn movement argued that lawns were not really about nature or beauty but an ostentatious display of wealth. In fact, so out of control were lawns in the United States that by acreage, lawns represented the largest irrigated crop—even more than corn. Something was seriously wrong. Maintaining pristine lawns involved sustained and large chemical inputs—fertilizers and herbicides—along with loud and destructive machinery. This is precisely what Rachel Carson warned in *Silent Spring*, required high school reading in Waisley. Students railed against the militaristic and violent rhetoric of weed control. Lawns were a form of grass chauvinism, an act of human domination over nature and must be opposed. Rather than acres of endless monoculture of turf grass, why not celebrate the planet's bountiful life?

More importantly, they asked, What is a weed anyway except a plant in the wrong place? Are settler colonists invasive species too? An animated discussion on the Chirp-Net deliberated whether we should extend the pejorative language of plants into humans. Drawing the slogan from *The Dandelion Insurrection*, "When fear is used to control, love is how we rebel,"[1] Chirp-Net concluded that it was important to move away from the language of hate and into one of love. They began a campaign to explore life in home gardens. While in Waisley it took the form of anti-lawn protests, other issues emerged elsewhere. For some it was saving a beloved species, challenging xenophobia, welcoming refugees, or collaborations with children across nations. Let us not promote a litmus test of geography, they concluded, but explore what organisms do and can do, and understand how we are all entangled in each other's lives. Children took careful notes and posted them.

Two other campaigns worked alongside the anti-lawn campaign. First was Love the Dandelion. The dandelion was chosen because it was ubiquitous. Thanks to British colonialism, the plant was everywhere. It emerged as a particularly worthy subject on the Chirp-Net. Across the world, the amazing properties of dandelions were celebrated. Dandelions are great for pollinators. In early spring, when there is not much insect food around, dandelions feed insects. The plants' leaves, flowers, and roots were made into delicious edible and nutritious food, tea, beer, and wine. Each day, new medicinal properties emerged—helping digestion and kidney function, improving immunity, and detoxifying the liver and gallbladder. Unlike the decreasing yields of commercial agriculture, dandelions were hardy, growing in many kinds of habitats. Now that there were no longer chemicals in the gardens, the plants were safe to eat. Like its relative the lettuce, you could harvest young leaves from the plant while the rest continued to grow.

Children in Waisley began cultivating multispecies gardens instead of lawns. Dandelions were a favorite. Other edible species, including some invasive and fast-growing species, were added. Global entries on the Chirp-Net revealed that one region's invasive species was another region's delicacy. If we all grew our palettes, our culinary habits could help our ecological problems.

Children exchanged recipes passionately and went on adventures, hunting for plants on roadsides, in abandoned lots, and in other degraded landscapes, where such species thrive. They enthusiastically ate countless new recipes: dandelion scrambled eggs with a garlic mustard pesto, dandelion salad greens, dandelion paneer masala, dandelion beet bortsch, knotweed curry, kudzu quiche, apple and knotweed pie. It was transformative. Children learned that a great deal of civilizational diversity and colonial inequality had been hidden from them. In the United States, urban children discovered the long botanical histories of numerous Native American cultures—for example, the Iroquois legend of the three sisters. According to legend, three crops—corns, beans, and squash—were inseparable sisters who grew and thrived together. In practice, corn acts as a pole for beans, beans stabilize corn and fix nitrogen in the soil, and squash, with their large leaves, protect all three plants by keeping the soil cool, moist, and weed free. Three Sisters Gardens were added into edible home gardens. These alternate models reminded children that plant life isn't about competition, a vision of nature red in tooth and claw. Plants also coexisted, engaging in mutualistic and symbiotic relationships.

Children embraced this vision. They brought international tastes into the household. They helped research, garden, cook, and innovate. Suddenly the whole household was involved—sharing the outdoors as active and engaged members. Parents, wanting to encourage their children, joined in. In many neighborhoods, these activities improved the health of the whole family. Global inequality had created problems the world over—too much food or too little. The Chirp-Net taught families about sustaining plant life, and food and medicine merged into a new philosophy of eating and health.

The sudden proliferation of diverse plants in front yards instead of chemically proofed lawns proved a bonanza for flora and fauna. This led to the second powerful movement, Migration Corridors. In many areas, children discovered that early hosts for endangered butterflies were critical to their survival. As children included these species in their gardens, towns were filled with butterflies attracted to the delicious nectar. Front yards became pollinating grounds for honeybees, bumblebees, butterflies, beetles, moths, and birds. Children discovered new ecological relationships. Through the

Chirp-Net they created temporal maps so that butterflies and other organisms could always find a Migration Corridor, where they found sustenance as they moved.

Of course, this was not always easy. Some neighborhood associations fought. Some households refused. But you did not need everyone, just enough. As the movement grew, its impact was worldwide, transforming not only the lawns and gardens but also the human lives and homes. From the small and cumulative actions of children who transformed their font yards and then their neighborhoods, communities, towns, states, and nations, a world was rebuilt.

INTERLUDE

A Cosmopolitan Botany: Tagore's Vision for Santiniketan

Visva Bharati represents India where she has her wealth of mind which is for all. Visva Bharati acknowledges India's obligation to offer to others the hospitality of her best culture and India's right to accept from others their best.

RABINDRANATH TAGORE

I have spent the last two decades following the xenophobic rhetoric of plant invasion biology. Invasive species are regarded as the number one problem contributing to the growing ecological devastation of the planet. In my work, I have argued that a singular focus on invasive species misplaces and displaces the symptom of the problem. We scapegoat invasive species for a problem that really arises from capitalism and greed. We have destroyed and overdeveloped our landscapes, leading to ecological devastation. Rather than address these underlying policies that continue to devastate the planet, we blame the foreignness of invasive species. Ever since I became aware of this, I have been drawn to alternate visions of ecologies, ones less nativist and ones that celebrate dynamic ecologies of relationalities. Rabindranath Tagore offers one such vision.

Tagore, a Nobel laureate, writer, poet, artist, educator, thinker, and painter, a touring figure regarded as a polymath and "Renaissance man," began a powerful experiment. He inaugurated an educational institution, Visva-Bharati, in Santiniketan; it opened in 1901 with just five students.[1] His was a global vision of egalitarianism, internationalism, interdisciplinarity, and cosmopolitanism: "I have in mind to make Santiniketan the connecting thread between India and the world."[2] While he began the institution as a Vedic school grounded in knowledge of the Upanishads and Vedas, he also drew on literature, poetry, and art from Europe. With time, it evolved into a much more broadly conceived sociocultural institution.

Tagore used his renown to establish this university in the rural community of Santiniketan. Visva-Bharati's motto was "Where the world meets in a single nest." All of Tagore's efforts—artistic, educational, and social—were informed by a universalist philosophy of self, society, and nature inspired by the Upanishads. As Tagore himself observed, "Education divorced from nature has brought untold harm to young children. The sense of isolation that is generated through the separation has caused great evil to mankind. The misfortune has been caused to the world since a long time. That is why I thought of creating a field which would facilitate contact with the world of nature." He encouraged students to feel the ground beneath them, asserting that the "soles of children's feet should not be deprived of their education, provided for them by nature, free of cost. Of all the limbs we have they are the best adapted for intimately knowing the earth by their touch. For the earth has her subtle modulations of contour which she only offers for the kiss of her true lovers—the feet."[3] In an essay titled "My School," Tagore wrote:

> Knowing something of the natural school which Nature herself supplies to all her creatures, I chose a delightful spot and used to hold my classes under big shady tree. I taught them all I could. I played with them. . . . We have there the open beauty of the sky, and the different seasons revolve before our eyes in all the magnificence of their colour. Through this perfect touch with nature we took the opportunity of instituting festivals of the seasons.[4]

Widely traveled, Tagore strove for a Pan-Asian unity that could meet the West on equal grounds. His was not a vision of exclusion—Visva-Bharati included the study not only of the Vedas and the Puranas but also of Buddhism, Jainism, Islam, Sikhism, and Zoroastrianism, as well as studies of world cultures, languages, and religions, including the traditions of the West.[5] His embrace of the Vedas and Upanishads was based on a principle of equal respect for all cultures, a prolific feasting of wisdom from varied cultures. Tagore's philosophy facilitated unity between all creation, including harmony between peoples across the world and between humanity and the natural world.[6] One can see this spirit of internationalism not only in its educational vision but in its gardens, in the trees in the area, on the campus, and in the blurring distinction between the inside and the outside.

I begin with a brief description of Tagore's vision of "universal humanism" and how it connects to his vision of the natural world, and then I contrast his vision with contemporary discourses on invasive species because Tagore offers a counterpoint. While grounded in elite privilege, Tagore's cosmopolitanism was grounded in an antinationalism, a refusal of borders for humans and animals. I conclude with some reflections about a universal humanism of the natural world.

TAGORE AND HIS UNIVERSAL HUMANISM

Reading Tagore's work in the midst of the rise of a global Hindutva is a singular contrast in vision. The caste privilege of the Brahmin Tagore is clear in his knowledge of the Vedas and Upanishads. Yet what a contrasting worldview! Scholars have argued that Tagore's universalist philosophy is expansive and not easily defined, particularly because it defies categorization within the strictures of the prevailing political and artistic movements of its time.[7] This philosophy has several elements.

First, for Tagore, universal humanism was a collective process. In his words, the "Infinite Personality of Man is not to be achieved in single individuals, but in one grand harmony of all human races."

Second, his view of "unity in diversity" included a unity of all things irrespective of nation, culture, or religion. He sought common ground not only in democracy and social organization but also in sustainability and human relations with the surrounding natural world. Throughout his work, and certainly in Santiniketan, one finds a concerted effort to bring nature and culture into a cosmic whole.

Third, not only does his view eschew an easy separation between nature and culture, it insists on an interconnected natureculture, but his is a decidedly elemental vision. In his essay "Tapovan" (Forest of Purity), Tagore writes: "The culture that has arisen from the forest has been influenced by the diverse processes of renewal of life, which are always at play in the forest, varying from species to species, from season to season, in sight and sound and smell. The unifying principle of life in diversity, of democratic pluralism, thus became the principle of Indian civilisation."

Fourth, Tagore carved out a third space, a hybrid vision. But his vision is not a passive embrace of the world but an active and self-directed hybridity. As a colonial subject, he did not reject western ideas and ideals, and similarly he did not fetishize all things Indian and subaltern. Rather, his was a studied and cultivated hybridity. As he writes in a letter to Albert Einstein in 1913: "My religion is in the reconciliation of the Super-personal Man, the universal human spirit, in my own individual being."

Fifth, Tagore's universalism is deeply grounded in Hinduism, in particular the Upanishads. His universalism relies on a view of infinity that includes "all the diversities of the world." He sought to find the infinite within the finite and worked for the unity of all life on earth.

It bears repeating in our times that Tagore eschewed nationalism and warned, rather presciently, that a nationalistic India would redefine Indian society through an imperial paradigm. Indeed, it has.

TAGORE'S NATURE

Rabindranath Tagore's nature is not singular but derives from an "eco-ethical" philosophy of human living and sustainable development. Tagore considers nature and human life as integral parts of the single entity, the omniscient, omnipresent, ubiquitous (*sarbang khallidang*), attribute-free (*nirguna*) Brahman. His capacious and inclusive philosophy emphasizes symbiosis and balance between humans and all other aspects of the living world (plants, other living beings, the Earth, the atmosphere, and the rest of the universe) and between humans and the world beyond (*moksha*).[8] Nonhuman lives were important—animals, plants, and the elements (air, water, fire, earth).

Tagore's nature is a nature without borders. He recognized the divinity of all life and the need for humans to live in harmony with the world. His love of nature included a romance with the rural over the urban. In 1862 Tagore's father, Debendranath Tagore, came across the land on which Santiniketan now stands. Enchanted with the site, he sat down for an evening meditation and was entranced. He subsequently bought the land.

Rabindranath Tagore's gender, caste, and class privilege are apparent in his writings and in his life experiences. For example, he recounts his first experience on the land:

> It was evening when we reached Bolpur. As I got into the palanquin I closed my eyes. I wanted to preserve the whole of the wonderful vision to be unfolded before my waking eyes in the morning. The freshness of the experience would be spoilt, I feared, by incomplete glimpses caught in the vagueness of the dark. When I woke at dawn my heart was thrilling tremulously as I stepped outside. . . . In the hollows of the sandy soil the rainwater had ploughed deep furrows, carving out miniature mountain ranges full of red gravel and pebbles of various shapes through which ran tiny streams revealing the geography of Lilliput.

Being transported in a palanquin to new property acquisition signals his upper-class status. Satyendra Basu observes that "when Maharishi Debendranath Tagore established Santiniketan the whole area is said to have been devoid of trees of any kind except in the few villages here and there which must have contained the usual type of fruit and bamboo groves etc. We can take it, therefore, that much of the greater part of the Flora now in Santiniketan and its neighbourhood owes its origin to the hand of man."[9] Basu notes that the soil was substandard quality. The topsoil had been eroded, often exposing the subsoil. In short, Santiniketan in the mid-nineteenth century was "a weather-beaten eroded barren place. In order to enhance vegetation and revive the greenery, the top layer of coarse grainy arid earth was removed and filled with rich soil brought from outside. Such a deep concern for the environment and zest for revival of greenery should indeed serve as a lesson for future generations."

From this entirely manufactured rejuvenated land, Debendranath Tagore created a global landscape. Yet in his eyes, it was all nature. No purity politics shape his view of natural and unnatural, tended or untended, pristine or cultivated lands. Along with his son Rathindranath, a horticulturalist, the elder Tagore realized a global garden at the university, a landscape in keeping with the international students and scholars who visited and the diverse and eclectic intellectual curriculum and philosophy it cultivated. For example, in the gardens of Uttarayan and in the surrounding area, he planted "exotic plants and trees from other lands. The African Tulip (*Spathodea campanulata*) from Equatorial Africa, the Sausage tree (*Kigelia africana*) and Rhodesian Wistaria (*Balusanthus speciosus*) from Tropical Africa, the Baobab tree (*Adansonia digitata*) from Sub-Saharan Africa and the Caribbean Trumpet tree (*Tabebuia aura*) from Latin America are some of the trees that have survived in Santiniketan as have the ideas and research studies done by foreign scholars who came to Santiniketan."[10]

The architecture likewise is synthetic and syncretic. The building Udayan is a synthesis of Indian, Japanese, Javanese, and European designs, finding commonalities in the traditions through abstraction and modern materials. Shyamali draws from a variety of influences where the design blurs the boundaries between indoors and outdoors by using the natural material of mud. Again, like Tagore's eclectic philosophy, the architecture of Santiniketan is not easily classified. Some have called his style a "Contextual Modernism," a new narrative for early modernist art in India, one that sees revivalism and modernist expressionism as part of the same goal: to open doors to aspiring Indian artists, freeing them from imperial restrictions and allowing them to give new life to Indian art.[11]

Tagore's Visva-Bharati at Santiniketan still has the rural trappings that Tagore dreamed of. Old buildings, even those made of mud walls and thatched roofs, still remain intact.[12] Indeed, by recent accounts, Tagore's global gardens have endured. Recently, the university sought the help of the Consul General of Japan to invite an expert gardener from Japan to restore the campus's three Japanese gardens. Alas, as Hindu nationalists have installed new leadership at the university, nativism and the "tinge of saffron" are beginning to undo this global garden.[13]

No doubt, Tagore's class and caste privilege are evident in many of his views. His grandfather, Dwarkanath Tagore, did business with the British, and so the family fortunes benefited from colonialism. One should understand Tagore's Bengali cosmopolitanism within this history. Yet it is striking how much Tagore exhibits an openness of spirit and vision, where people, plants, animals, literature, arts, and sciences are welcomed from around the globe. While opposed to nationalism, Tagore shows a deep reverence for Hindu philosophy and also rural life. As he writes, "The ideal of perfection preached by the forest dwellers of ancient India runs through the heart of our classical literature and still dominates our mind."[14] There is a profound romance of nature, even if it is not nativist: "In our dreams, nature stands in her own right, proving that she has her great function, to impart the peace of the eternal to human emotions."[15]

Tagore was an internationalist and was decidedly anticolonial. He critiqued the sameness of the mechanistic nature of western industrialism in contrast to the harmonious diversity in the forests of the east.

CONCLUSION

Before biology invented the category of invasive species, other imaginations of spatial geographies of plants thrived, Tagore's vision being one of them. Indeed, colonists also exhibited an enthusiasm for vegetal adventure with a range of plants, animals, and pathogens. Colonists too globalized the planet by moving species. Horticultural societies, repositories, gardens, and herbaria across the West cultivated the exotic and curious. Yet these practices were part of a politics of domination, acquisition, consumerism, and conquest. Tagore presents a contrasting vision. To him, accepting plants and traditions from across the globe are not about conquest or appropriation but a welcoming of multicultural and multispecies ecologies. In *Religion of the Forest*, he extolls the need to establish relations with the universe through union, through sympathy rather than the cultivation of power.

> According to the true Indian view, our consciousness of the world, merely as the sum total of things that exist, and as governed by laws, is imperfect. But it is perfect when our consciousness realizes all things as spiritually one with it, and therefore capable of giving us joy. For us the highest purpose of this world is not merely living in it, knowing it and making use of it, but realizing our own selves in it through expansion of sympathy; not alienating ourselves from it and dominating it, but comprehending and uniting it with ourselves in perfect union.[16]

This was a different vision of India. With the rise of Hindu nationalism in India today, Tagore's critique of nationalism is prescient. Far from an insular world that looks inward and to the past, Tagore welcomed the outside, the alien, and the foreign as partners for more just futures.

PART FIVE

Uprootings

You must resist the common urge toward the comforting narrative of divine law, toward fairy tales that imply some irrepressible justice. The enslaved were not bricks in your road, and their lives were not chapters in your redemptive history. They were people turned to fuel for the American machine. Enslavement was not destined to end, and it is wrong to claim our present circumstance—no matter how improved—as the redemption for the lives of people who never asked for the posthumous, untouchable glory of dying for their children. Our triumphs can never compensate for this.

TA-NEHISI COATES | *Between the World and Me*

CHAPTER NINE

Vegetal Sublimations

Cartographies for Adisciplinary Sciences

How to tell a shattered story?
By slowly becoming everybody. No.
By solely becoming everything.
ARUNDHATI ROY | *Ministry of Utmost Unhappiness*

I change myself, I change the world.
GLORIA ANZALDÚA

I BEGAN THIS BOOK THROUGH THE SUBLIME. Through an exploration of the coloniality of botany, I argued that the sublime in science has rested on a primacy of human worlds and a separation of human and nonhuman worlds. The imaginations of decolonization from feminists, indigenous, postcolonial, decolonial, and queer activists all demand the rejoining of human and nonhuman into more-than-human worlds. How? If we abandon our botanical amnesia and rememory the history of botany, the embrangled past opens up. These histories show us how and why academic disciplines and subdisciplines, developed and consolidated through colonialism, have produced structures of coloniality—nomenclatures, taxonomies, epistemologies, methodologies, methods, ontologies, and theories sanctified by liberal logics. The original colonial bioinvasion is forgotten in invasion biology. The Linnaean "marriage of plants" helped usher in a heterosexual world promoting monogamy, and the *Hortus* laid the conditions for modern biopiracy.

Why reimagine botany? Science is in many places—in bacteria, plants, and animals.[1] While science can occur in many sites, such as indigenous knowledge systems, kitchen science, home gardens, or DIY science, my work seeks to open up the epistemic authority and expand the practices and epistemologies of what we call "normal science." Some may find this a futile task; indeed, Audre Lorde long warned us about the limited conceptual apparatus of academe, and how the master's tools can never help us dismantle the master's house.[2] But three decades of feminist STS have also revealed that there are no sites of purity in the world. Within the workings of botany are not only histories of colonial domination but also subaltern knowledges.

There is a deep resonance of theories across the humanities and sciences. Most academic disciplinary formations owe their development to the growth, expansion, and machinations of colonialism. It is hardly accidental that there are many parallels and synergistic evolutions across disciplinary formations. For example, Karen Cardozo and I have argued that the idea of genera in biology is intimately linked to the idea of genre in literature.[3] We enumerate these shared histories and conceptual tensions as deeply resonant and parallel formations. Our world is deeply embrangled, and we need to understand where the coevolutions and synergies lie. If we fall back on our disciplinary silos, we perpetuate "normal" science and its processes—climate change, overdevelopment, ecological devastation, environmental crises, energy dependence, poverty, reproductive injustice, unsustainable agriculture, and food insecurity, to name just a few. Seeking to understand the world as embrangled in its histories is the urgent project before us.

Methodologies of the Pressed

The herbarium, a global mortuary filled with dead dried plants, is perhaps an apt metaphor for empire. How does one theorize a world of pressed plants? I began and have engaged throughout with the thinking of feminist, queer, indigenous, postcolonial, antiracist, and anticolonial scholars and activists; I wish to end with them as well. I am drawn to alternate paradigms grounded in a refusal of colonial modes of masculinist inquiry, preferring instead to model relations and relationalities built on justice. Chela Sandoval offers love as an opening for social justice. Patricia Williams urges us to work with vulnerability.[4] Instead of the hypermasculinity of disciplines and learning to "fake it till we make it," we can make spaces for living, where vulnerability is not weakness. In this era of climate change, to imagine being connected to one another, owing each other something, and recognizing our vulnerability to injury and redress are critical for difficult conversations about decolonization. Artist Shahzia Sikander offers us intimacy: "Ultimately I want to interrogate what it means to decolonise. The conversation is so male-centric, and focused around retribution and erasure. Could we conceive of decolonisation as something else? Perhaps reframe it in terms of intimacy?" she suggests.[5] Kim TallBear offers "being in good relation."[6] In critiquing the grand narrative of American exceptionalism, she offers relational webs as a spatial metaphor for attending to our relations and obligations in the here and now. Together these scholars offer feminist frameworks for our actions and ambitions—love, vulnerability, intimacy, and good relations.

Chela Sandoval's influential *Methodology of the Oppressed*, which inspires this section's title, marks one such intervention.[7] Sandoval offers a robust methodological framework to reveal how US third-world feminisms have altered our perspectives on contemporary culture and subjectivity. By exploring the development of an oppositional consciousness that mobilizes love as a category of critical analysis, Sandoval opens new possibilities of alliances and politics. These works, along with others from feminist, Black, Latina, Asian, indigenous, queer, immigrant, and postcolonial studies, have deeply inspired me. They offer tools to move beyond a decentering of normative structures of western academe to new epistemes grounded in the alterity of empire. Here I suggest that we extend them to develop methodological innovations for multispecies worlds.

Two particular insights inspire. First, we reject binaries between imperial and subaltern knowledges because the master's tools were often "fashioned by subalterns—whose social location and political desires left imprints on the tools themselves."[8] We need rich accounts that trace the dense imbrications of indigenous knowledges *within* the canon of western and imperial science.[9] Second, Leanne Simpson reminds us that a profound legacy of western imperialism is to remove agency from plant and animal worlds and to transform them into natural resources.[10] How do we then develop methodologies of the pressed that accounts for the violence and appropriative ambitions of empire?

For Sandoval, the answer is "love"—not the love of romance novels but love reinvented as a political apparatus, retooled as a potent technology that brings together diverse knowledge practices that can help rebuild the world.

In drawing on Sandoval and extending her "differential consciousness" of third-world women into the vegetal world, I am not suggesting some naive alliance or symmetry between the oppressed worlds of the human and vegetal. Rather, I mean to extend the analytic power of Sandoval's love to understand plant words as agentic worlds, not like human but in their own right. Sandoval's work engages questions of difference across a plentiful life on earth. The problem is not difference; as Audre Lorde has long reminded us, "It is not those differences between us that are separating us. It is rather our refusal to recognize those differences, and to examine the distortions which result from our misnaming them and their effects upon human behavior and expectations."[11] If we extend Sandoval's, Lorde's, and Simpson's challenge to recognize colonialism as an ecological project, we understand how imperial ambitions set out to objectify and commodify not only some people but also animal and plant worlds. With this understanding, we enter a rich terrain of analysis. Rather than being reduced to their ancestors on pressed herbarium sheets, plants emerge as agentic, agile, and transforming organisms. They mutate, adapt, evolve at every turn in "tree time."[12] We should not rush the vegetal world. Imperial sciences and their rapacious desires tried to speed up plant time through control and through sciences such as plant breeding, conservation biology, restoration ecology, industrialized agriculture, and invasion biology. We can and must do better.

By moving away from human-centered worlds, we open dizzying new conceptual landscapes. We need to reckon with both anthropomorphism and human exceptionalism. Anthropomorphism projects human frameworks onto nonhuman worlds. Conversely, if we resolutely refuse to use human frameworks to understand nonhuman worlds, we are in danger of practicing exceptionalism by denying that sentience, agency, and emotions belong to the nonhuman. The dual risks and challenges of anthropomorphism and human exceptionalism invite deeper consideration. This is rich territory.

In queering our tales of life on earth, we must, as Eve Tuck reminds us, be wary of the dangers of "damage-centered research" that obsessively documents pain and loss, a "pathologizing approach in which oppression singularly defines a community."[13] Instead, she suggests we move to a desire-based framework—where we don't fetishize damage and degeneration but celebrate survival and "regeneration." We can queer our understanding of ecologies by embracing the many ideas that scholars and activists have offered. These accounts inspire lush cosmologies, long ignored and erased. Often just footnotes, such works should foreground our syllabi. We must, for example:

Foster "care" (de la Bellacasa) and "caretaking relations" (TallBear) for all "earth beings" (de la Cadena).[14]

Reject "animacy hierarchies" (Chen, TallBear) and "white empiricism" (Prescod-Weinstein).[15]

Develop alternate epistemologies such as a "feeling for the organism" (Keller) or a "feeling around the organism" (Roy), and "agential realism" (Barad). We must, as (Myers) reminds us, unlearn detachment and instead learn to become "sensors" (Myers).[16]

Learn alternate modes of engaging with the world so as to "think like a forest" (Kohn), or process the colonial past by learning to "dream of sheep in Navajo Country" (Weisinger). We must learn to not refuse the messiness of history and always "stay with the trouble" that our histories bring (Haraway) and yearn for cosmic and scientific lessons in traditions such as "braiding sweetgrass" (Kimmerer).[17]

Embrace "affective ecologies" (Hustak and Myers) such as "black livingness" (McKittrick) and Matsutake mushroom worlds that defy neat borders and force us to deal with "unruly edges" of naturecultures (Tsing).[18]

Stop appropriating indigenous knowledges, and insist that "pollution is colonialism" (Liboiron) and "decolonization is not a metaphor" (Tuck and Wang).[19]

Queer our understandings of ecologies (Mortimer-Sandilands and Erickson), undo monogamy (Willey), and foster trans queer ecologies of justice (Wölfle Hazard).[20]

Engage in nonreductionist science by understanding ecology holistically, such as taking soil and underground life seriously (Lyons; de la Bellacasa).[21]

Eschew a politics of purity (Shotwell) and an idyllic and mythic past to work with damaged bodies and worlds. After all, we are already chemically impure as "alterlife" (Murphy). Frameworks such as "planetarity" highlight naturecultural worlds that exceed capitalist globalization (Spivak).[22]

Practice interdisciplinary work through the conceptual wisdom of literary scholars such as rememorying (Toni Morrison), vital ambivalence (Jamaica Kincaid), uses of the erotic (Audre Lorde), making familiar worlds unfamiliar (Suniti Namjoshi), and fostering speculative and radical science fiction (Octavia Butler, Ursula Le Guin).

Actively question the colonial roots of savior sciences (Parreñas, Chacko) and work to decolonize methodologies (Smith).[23] We need to recognize the breathtaking diversity of life on earth while resisting the deep impulses within science and technology studies for facile biological metaphors. The particularity of vegetal worlds means mushrooms aren't the same as grasses, grasses aren't like gingkoes, gingkoes aren't like tumbleweeds, tumbleweeds aren't like dandelions, dandelions aren't like eucalyptus, and eucalyptus aren't like oaks. Each has uniquely evolved biologies, and their varied evolutions and biological and social histories are firmly embedded in their spirited lives. For example, I am troubled by the effortless ways in which the term *rhizome* and the idea of the rhizomatic have taken root in STS. No doubt there are plant rhizomes, but suggesting that some essentialized organic form is shared across science and the humanities is problematic. The vegetal world is breathtakingly diverse. For every rhizome, there are nonrhizomes; for every mutualist, a parasite; for every cross-pollinator, a resolute inbreeder.

Yes, there is a biological exuberance of life, and indeed it is a stupendous imaginary to borrow from, but careful and contextual adaptations of individual species cannot and should not become naturalized and scientized into universal theories of the evolution of knowledge. Rather, recognizing the multiplicity and astonishing plurality present in plant worlds should foster a more engaged, interdisciplinary, contextualized, nuanced model of knowledge making. We need to move from *binary* logics to *multinary* ones, beyond clonal rhizomatic imaginations to the promiscuous possibilities of vertical and horizontal inheritance that the worlds of plants, bacteria, and viruses unleash. Bacteria and viruses that move genetic material across species challenge binary logics and even species logics.

And yet through all this multiplicity, it is worth remembering that all life on earth is still genetically remarkably similar.

Cartographies for Adisciplinary Sciences

How do we use insights from feminist, postcolonial, anticolonial, antiracist, and indigenous perspectives to "practice" experimental biology in ways that challenge the coloniality of science?

The heart of any project on decolonization must include dismantling colonial epistemologies of domination—where some humans claim dominion over other humans and the nonhuman worlds around them. Instead, we must work *with* and *alongside* other scholars and organisms on the planet. Scientists' claim to aperspectival knowledge—knowledge that is objective, neutral, and value-free and that produces universal "TRUTH"—is one we need give up in favor of more reflexive, historical, and contextual epistemologies and methodologies. In short, we should take decolonial, indigenous, and postcolonial critics seriously in developing more modest sciences that approach the complexities of the world with greater humility, and in producing situated and partial knowledges and "truths."[24]

We need to historicize biology, attending to the ghostly, spectral pasts and presents that haunt. In *Ghost Stories for Darwin*, I temper the usual progressive and liberatory vision of science to consider the harms of science, especially eugenics and its far-reaching impact. I use ghosts as a metaphor for the many bodies violated, sterilized, experimented upon, maimed, killed, and exterminated. I have found living in these haunted ghostly worlds sobering but also immensely clarifying and rewarding. Academic work matters; its consequences are all too real. Biological hierarchies are profound: not only are some lives deemed more important than others, but some are deemed more violable and dispensable than others. Women and people of color are rendered as more animal-like, some animals as more passive and vegetal, and some vegetal life as more nonsentient than other such life. Invoking Audre Lorde again, I assert that difference is never the problem. We can celebrate our differences—but when differences mingle with power, some groups dominate and oppress other groups. These ghostly worlds are filled with souls wounded by the violence of scientific domination. We cannot just put it behind us. In overcoming our botanical amnesia, rememory must be an active project of decolonization.

Decolonization is not a singular act but necessarily a continual process, needing constant tending, reflection, and action. Much like the thigmatropic tendrils of morning glories that have inspired so much of my work, the tendrils of colonialisms have tangled into dense thickets and knots and (inter) disciplinary formations that defy untangling. Given the long passage of coevolution, there is no place of originary return. Gender, race, sexuality, class, and caste may all be biological categories solidified through colonial politics of difference, but they have since tangled their way into new social formations of the present. Tracing these embranglements is important, but hopes of untangling them are futile, indeed impossible.

Agnotology is a field that focuses not on individual ignorance but on "culturally induced" ignorance—a systemic and cultural process that leads to socially sanctioned, willful acts of forgetting and systematic failures of memory.[25] In an academy that is organized into disciplinary silos, disciplines don't just engage in benign neglect but act as powerful and purposeful systems of willful forgetting. Interdisciplinary work is one way out of this.

In his foundational text, *How Europe Underdeveloped Africa*, Walter Rodney argues that Africa was deliberately exploited and underdeveloped.[26] In short, Africa developed Europe while Europe underdeveloped Africa. Colonialism does not only explain the ascension of Eurocentric knowledge, it also explains the erasure and marginalization of colonized knowledges through the deliberate and particular actions of colonialism. The growth of colonial sciences elevated Eurocentric knowledge at the expense of local and indigenous knowledges. To decolonize knowledges we must understand the decimation and erasure of local knowledges. This remains equally true in postcolonial nations that continue to privilege Eurocentric knowledges. Linking development of colonial powers alongside underdevelopment of the colonies is critical to our historical understanding as we seek projects of decolonization.

In choosing the science of botany as a site of reinvention and reimagination, I realize that I am picking a site fundamentally shaped by histories of colonialism. I am drawn to Jamaica Kincaid's ambivalence, critical for an immigrant gardener who wishes to eschew the purity politics of celebrating native species even when struggling with dominating plant species in her garden. Kincaid chronicles the constant ethical push and pull of "the relationship between gardening and conquest"[27]—to be a fierce and antagonistic critic of colonized domination even while a bourgeois gardener. Refusing purity politics, she draws on "decolonial ethics through split forms of self-representations that refuse modernity's insistence on a uniform self."[28]

Kincaid's vital ambivalence and the conceptual power of "impurity" propel me against purity politics—nativism, xenophobia, walls, isolation, and incarceration all align in a quest for a nostalgic return to an imaginary idyllic past. Simultaneously it also challenges the romantic politics from the left that seek unitary salvific "returns" to and claims to a future. In this context, a promiscuous, polymorphous, interdisciplinary methodological approach is precisely what we need to remind us that we are impure beings in impure nations with impure histories and genealogies. We should revel in impurity, embrace it as central to our politics, as methodologies and methods. As Mohsin Hamid reminds us, "In these pure times, you believe more impurity is desperately needed. Only impurity can save us now. But, fortunately, there are reasons for hope. Our species was built on impurity, and impurity will probably come to our rescue once again, if we let it."[29] It's time to reframe impurity for what it really is: embranglements of colonial power through the vital and vibrant life on earth.

CHAPTER TEN

Dreams of a Lively Planet

Persevering in one's existence is the particular quality of the organism; it is not a progress towards achievement, followed by stasis, which is the machine's mode, but an interactive, rhythmic, and unstable process, which constitutes an end in itself.
URSULA LE GUIN | *Dancing at the Edge of the World*

WHILE IT IS CUSTOMARY TO TALK about the resilience of the planet, this project has reminded me of the resilience of slavery and colonization in their powerful and ongoing afterlives. Decolonization may be impossible, but the impetus is necessary and vital. Decolonization is not a singular act. It is a pledge, a promise to overcome botanical amnesia, rememory the violence of colonialism, and rewrite, repent, and repair history and its material violence. After all, colonialism isn't something in the past but an ongoing and lasting process.

While I have focused on plants, we cannot talk about plants without talking about humans, life more generally, or the planet. The biggest lesson I have come away with is the impossibility of telling a story of plant life without the histories of colonialism. Yet my biological education succeeded supremely in doing just that.

Colonialism knits the world together, the natural, cultural, political, and planetary. It knits us all together into a planetary tapestry. Pull one thread, and it all unravels. As David Noble reminds us, "Each major scientific advance, while appearing to presage an entirely new society, attests rather to the vigor and resilience of the old order that produced it."[1] The science I was taught was singular and universal in its claims, and its focus was purposefully myopic. It shut out the world's diverse cultures, histories, and politics in the name of objectivity. Botany was indeed empire's botany, cultivating a science, shaping its infrastructure, and enlivening its cultures for the success of empire. These characteristics continue to shape knowledge systems today. The tragedy is in its distortions of plant life *and* its embrace of epistemologies that impede our ability to experience and learn from life. Science corrects itself, I'm often told. Not in this instance. Changes are so small and slow. The task before us is to recognize the illusion of objectivity for what it is and embrace more ambitious philosophies of meaning making and understanding. We urgently need to develop new epistemologies and actively work toward decolonization.

The Imperatives of Decolonizing

Colonization and decolonization projects grounded in colonial logics are not merely theoretical but are embedded in the coloniality of power that lives on in our scientific (and economic and political) infrastructures. For example, discussing our failed climate policies, Amitav Ghosh astutely notes:

> To look these facts in the face is to recognize that it is a grave error to imagine that the world is not preparing for the disrupted planet of the future. It's just that it's not preparing by taking mitigatory measures or by reducing emissions: instead, it is preparing for a new geopolitical struggle for dominance.[2]

Current climate policies are what the elites want, what the afterlives of colonialism demand: a voracious appetite for power and resources and wealth. The elite are convinced that they will survive, and the marginalized have long lived with the ravages of climate change. Nothing is new here, despite liberal talk of climate mitigation, technological innovation, or sustainable sciences. The price paid has been high for the marginalized. What we need is not equality but justice. As George Lipsitz warns about the recent turn toward inclusion and diversity rather than justice, "Desires for self-determination and dignity become channeled by the power structure into demands for roles inside oppressive systems rather than for changing those systems themselves."[3] A radical transformation is demanded of us.

The literature is studded with wisdom. Samir Amin's work warns us that colonization was about material resources and thus that decolonization cannot be accomplished through changes in epistemology alone.[4] "You can't decolonize the curriculum without decolonizing the world—we still live in a colonial world," Robin Kelley reminds us. No project of decolonization can be launched from the ivory towers of history. Faye Harrison stresses that for a radical transformation we need decolonization *and* democratization.[5] If we are serious about decolonizing, we need to clean house and move out of our ivory towers and onto the streets. We need to build alliances, forge collaborations, and strengthen communities. For some of us, this work requires learning to speak; for others, it requires us to listen, stay on the sidelines, and make space for long-silenced voices. But many questions remain. As Tania Pérez-Bustos asks:

> how homogeneous is the idea of decoloniality being used in Northern contexts? What kind of systematic ignorance accompanies this homogeneity? Whose singularities are being lost in terms of theory? Why? Is the use of decoloniality, or better yet the search for decoloniality, decolonial enough? Decolonial in what sense? Or is this search for decoloniality actually reproducing certain geopolitics of knowledge and logics of colonialism? From my experience as a feminist STS scholar based in Colombia and not representing anyone, with my singular voice, I would say it might be.[6]

In their now canonical essay, "Decolonization Is Not a Metaphor," Eve Tuck and Wayne Yang warn of the increasing depoliticization and domestication of the term *decolonization*. It has become a buzzword, superficially adopted into education and the social sciences such that everyone is attempting to decolonize everything, but in fact "it recenters whiteness, it resettles theory, and extends innocence to the settler, it entertains a settler future."[7] Decolonization necessitates material practices, rematriating land, and actively engaging in reparative and restorative practices.[8] As we embark on the impossible but necessary project of decolonization, moving beyond metaphors into the material impact of colonialism is critical. We must be wary of narratives of progress, a warning Priya Satia offered earlier in the book. Often decolonization, as well as calls for justice and reparations, are articulated in the language of "progress" and being on the "right side of history." It would behoove us to refuse progress narratives. Rememorying planetary life is about life constantly remaking life. In releasing the Linnaean thread, we must give up our amnesia and embrace the resplendent rememories of our embrangled life on earth.

Fables for the Mis-Anthropocene: The Memory Gardens

The sun gently rises over the horizon. Slowly it brightens the expansive landscape, casting its warm spell on a land reconfigured. This is New Pangaea. This is the gift of prophecy.

Generations after Chirp-Net was launched, the descendants of community members were ready; they had carefully planned and prepared for this time, filled with excitement and anticipation. The planet—lands, oceans, and bountiful inhabitants—were together again. The descendants had sharpened their senses, retooled old technologies for their new needs and new ways. If not for the Chirp-Net, who would have known these unusual genealogies? Billions of years earlier, continents on earth had begun as a single land, Pangaea, surrounded by the Panthalassan Ocean. In Pangaea, life blossomed across oceans, airs, and lands. The earth's mantle came alive in a gentle simmer. It burped, it belched, it hiccupped, it gushed, it spewed, it spouted, it spurted. Its ferment sent cascades of tremulous waves, setting land masses afloat to meander and circle, supple and flexuous. The continents danced in lazy circles, jostled and touched each other in playful pirouettes, floated across and cruised the poles and tropics. Some land masses said goodbye; other new ones appeared. Some species were separated, others united. Alien species befriended one another, creating new webs of belonging. The earth simmered more, producing new friends, new lovers, new mates. The planet kept moving into new formations of land and ocean. Extinctions, births, life, and death arose in an endless cycle. New creatures emerged, new food, friends, foes, and lovers. A world that had drifted apart three hundred million years earlier was now back together after billions of years of adventures, a cornucopia of vital life and living. Who are kith and kin now?

Time blossomed and spawned many temporal progenies. The story is told in so many ways. Do we follow the time of the universe, the time of galaxies, planetary time? The geologic time of rocks, an endless, expansive eternity? For them it was Pangaea last year, the mighty Everest last week, the erupted volcano a minute ago, massive fires burning a second ago. Whose time is earth time? The endoliths live for millions of years, with a generation lasting ten thousand years. Sponges, corals, and tubeworms live for tens of thousands of years. Some redwoods, yews, olives, figs, cypresses, and pines last for thousands of years. Some sharks, tortoises, whales, whales, rockfish, and sea urchins survive for hundreds of years. Some creatures live tens of years and others a few years, minutes, even seconds. Whose time is earth time? For centuries humans have made it about human time. What an impoverished view of life on earth! What did humans not see? What can they not see?

Pangaeans decided to give up human time, to revel in the multiplicities of time and temporalities. Pangaeans inhabit many times, planetary time as large as the expanse of planetons and as tiny as the fleeting zeptoseconds. They inhabit multiple timescapes, speeding up, slowing down, expanding, contracting. It is thus that they came to be anointed as the planet's memory keepers. They hid in different time zones, their presence fleeting and unrecorded in the histories of vital earth. As they chronicle the new botanica, they chronicle a bountiful earth. Why should there be only one story of the planet? Let us tell many stories, even contradictory ones. Why does there need to be coherence? Now that New Pangaea has come together, memory keepers have taken form, become visible, a vital community among a teeming life on earth. Each day, they wander the planet, their sensoria in full alert, taking in the vitalism of the newly conjoined landmass. With joy, play, and curiosity, they wander the planet and become acquainted with their new brethren. Who are they? What do they know? Where do they come from, and where do they wish to go?

Sura, one of the memory keepers, wakes up and puts her hands to the ground. She feels the humming pulse of the planet, the gentle rumblings of pleasure, the quiver of recognition, the reverberations of histories lost and found. She wanders through the dense thickets collecting fragmentary thoughts and memories from the world around her. For some species who were born in the last minute, the last day, the last week, this is the life as they have always known it. Nothing seems different. Others wander around in shock, traumatized by landmasses colliding with each other. Their homes have been lost, their loved ones have disappeared. Mountains appear where none existed. Oceans churn where there was no water. Sura absorbs, taking it all in.

Zuri, another memory keeper, stayed low to the ground, sensing the busy life underground. They hear the roots finding their way across new ground, finding new roots, connecting, entangling, sensing old kin and new. They sense the insects, worms, and small mammals scurrying and wriggling around, feeling their way in and under the soil. The soil itself is vital and vibrant, sending messages from roots to the overstory as vegetal life appraised its new world. The world being remade yet again. The sun's ever-stable presence is comforting, refreshing old routines. Organisms search for water and minerals. This is new work. There are new worlds to be discovered and life to be made anew.

Tiva flies, soaring among and above the tree canopies, listening to the birds sing to each other. New songs are composed, new melodies learned. The planet's atmosphere, the air, is one contiguous space and yet, like the ground and soil underneath trees, inextricably interconnected with the plants, animals, and birds. The planet is entangled. Birds of prey soar, surprised at new creatures they have never seen before. Tiva flies into the overstory, enjoying the changing patterns of light and dark as the sun travels the skies, as the world comes together yet again.

Essi roams among the sweet potatoes. Long-lost relatives who traveled the oceans and prospered in new lands are back together. So many of them! Ah, the humans loved sweet potatoes and propagated them endlessly across the planet. She watches amused as the tubers send messages to each other sharing their joyous exploits, painful losses, and tales of the lands they have visited. Thanks to humans on the planet, some clones have been cultivated for thousands of years, storing their ancestors' memories in their cells. *Buddy*, Essi heard one tuber tell another, *you made it all across the ocean to Polynesia that long ago? Wow! Tell me more!*

Keya wanders deep into the wet, lush forest to the land of the orchids. Orchids stare at each other. *Why did you develop that red spot?* one asks another. *Ah, my mate is the red-spotted butterfly. Oh,* responded the first, *mine is the blue-tongued beetle.* They revel in their kin and their bountiful variation. They regale each other with sensuous tales of pleasure and desire with their insect lovers. Back together again in a new land, new innovations have been divined and new evolutions await.

Elu and Maji hover above the ground, scanning the planet, chronicling evolution's lost and found. The many evolutions in exoskeletons, limbs, sensoria—almost no species are the same. Through millions of years, they have adapted to changing worlds. Elu and Maji can hear the varied tempo of the many lives. The momentous ice ages, the droughts, monsoons, storms, and tsunamis, the warming planet, the melting glaciers. The unpredictable decades with climate change. Thankfully, most of the rapacious humans died off, unable to sustain themselves. Bending deep, wanderers can feel it—that moment in time when the rapacious humans gave up and the planet was able to breathe again. They can see the joyous innovations that exploded when life did not have to revolve around the profits of some greedy humans.

Kai slowly expands its body, transforming into a juicy blob of luscious protoplasm, a magnet for bacteria and viruses. As they enter its body, it bursts with joyous laughter. It senses DNA from all over the world, a kaleidoscope of patterns, sizes, and shapes. *What a khichidi,* Kai says, *a mishmash of global DNA. They are truly the cupids of the planet,* Kai thinks. Here in their tiny, infinitesimal bodies lies the story of life on earth.

Elu has been waiting for this moment. It slows down time. It has long been waiting to see the majestic Garden of Eluru again. How did it fare through the eons? It was once a luxurious garden, an envy of the planet. Elu approaches the garden and is stunned. It is unrecognizable; nothing is the same. Elu examines it closely. It is a brilliant landscape, still luscious and luxurious, still plentiful, still beautiful. Colors explode, trees sway, flowers bloom. Life scurries about, birds sing, insects chirp. Elu thinks, *Evolution is about loss, evolution is about change.* Eluru is a ruined garden to some, a resplendent beauty to others.

The future that awaits us offers not transnational but un-national, even anti-national ecologies. The joy of planetary life is in the now, for the now. In the living, for all the living. Extinctions, perhaps, but no apocalypse. Until, of course, the sun dies. But there are always new suns.[1]

INTERLUDE

Against Eden: Futures Lost and Found

EDEN

They dreamed of Eden, an imagined land—clean, bright, luxuriant. Eden framed their civilizational stories—of good and evil, temptation and greed, trees of knowledge and life. Edenic philosophy beckoned them to conquer—land, plants, animals, people, indeed the world—proselytizing and imposing an Edenic planetary regime. Edenic science—its dominion over Absolute Truth, value-free knowledge, and objectivity—expanded and consolidated Eden into a superior settler science, a colonial science, and a global science. Through colonialism, slavery, and conquest, Edenic science was propagated. Edenic science was a eugenic science, classifying the world into inferior and superior bodies. Only the best were acceptable for Eden—the whitest, richest, most elite, masculine, beautiful European subjects. Edenic science was manifest destiny.

It did not end well. Eden soldiered on, and at the brink of apocalypse, Edenic science was exposed as a mirage conjured by false gods, mendacious prophets, specious truths, spurious ideologies, and distorted knowledge. Greed, money, inhumanity, and short-term gain had ravaged the planet. They nearly extinguished all life on earth. But we, the survivors of a group that have long challenged Edenic philosophies, were ready. We were waiting for this moment; around the globe, the vanquished rose up against the hegemony of Edenic science, its myths, mythologies, and fictional facts. As multispecies communities, we nurtured ecosystems that could revive the planet. We began anew, establishing an Anti-Eden.

ANTI-EDEN

For Anti-Eden, we rejected the usual story of "two cultures"—humanities and sciences. Lost in such a genealogy are women of color feminists, indigenous feminisms, and postcolonial, diasporic, and queer feminists, who have always written more syncretic, symbiotic stories that do not center the Euro-human, privilege the "human," or separate the humanities and the sciences. We built a world with their vision. There is no purity, no Eden to return to. We renewed our imagination, traveled paths not taken, revived many futures that had been denied. Not all these ideas were new; they had circulated for centuries, ignored, ridiculed, discarded. But this is our path now—lost futures, found and reimagined into more just worlds. Anti-Eden is guided by Anti-Edenic principles.

ANTI-UNIVERSALISM

Edenic science enforced a universal Linnaean taxonomy for all planetary life. It erased local knowledges and naming and replaced them with a universal binomial system of hierarchical nomenclature in Latin, a tongue foreign to much of the world. Edenic science disenchanted the world, erasing luxurious and magical contexts of countless meaning-making systems and severing plants and animals from local knowing. For many who would never travel the globe, Edenic science was a study in alienation from the natural world.

We resisted the colonial project of universality. We resurrected and reclaimed local knowledges, practices, and wisdom. Names and knowing from across the world held a cornucopia of generational wisdom—worlds within worlds, myriad ways in which cultures experienced, knew, and imagined the intimacies of life and living. We discovered new relationalities, new kinships, new worlds. We used technology to invent a "universal translator," allowing us to listen and share across worlds. Though it was never perfect and much was lost in translation, we celebrated patience, negotiation, and compromise. With time, all languages added new words, new meanings, new relationalities, and new imaginations to their lexica and worlds.

ANTI-SPECIES

Taraxacum officinale was, according to taxonomists, a single species. But was it? The scientific term obscures its polysemic and engaged life and many roles: dandelion, tarashaquq, lion's tooth, blowball, faceclock, pee-a-bed, pisacan, swine's snout, warm rose, butter flower, milk bin, lootari, payasvini. While some humans named it a pest and poisoned the earth to get rid of it, others reveled in its presence, endlessly bountiful—delicious to the tongue, cure for the liver, deployed as a diuretic, laxative, wart remover, and insect repellant. *Taraxacum officinale* is not singular but multiple and endlessly generative. The concept of species, we concluded, obfuscates rather than enlightens. We are anti-species.

ANTI-INDIVIDUAL

Edenic science is grounded in the individual as the unit of ecology and evolution. Which individual is fittest? How do we measure reproductive success? But health, we believe, is not singular but contextual, environmentally grounded, entangled in multispecies and community ecologies. We discovered that some species losses caused the loss of many others. Long-ignored soil and microbe communities sustained and shaped much of life. Everything matters: the health of communities, relationality, kinship, love, caring, the give and take of life, the changing seasons, the waxing and waning of the moon, the ever-changing and evolving landscapes. We have new vocabularies that help tie our worlds together. We measure life and health not as individuals but as contextual collectives. When one is lost, so much more is lost and gained. We have to think beyond one. We are anti-individual.

ANTI-BIODIVERSITY

Numbers and counting were at the core of Edenic science—a singular obsession with figures, metrics, and statistics: 110 species here, 3000 there. Always numbers. The concept of biodiversity poorly captured what we valued about the world. Edenic science ushered in a world of extreme climate change. Today, none are well adapted to the worlds they live in—origin stories, nativism, and nostalgia are useless concepts. We need to live in the ecologies of today, build new relations, new intersectional community ecologies. We must stop using diversity as some magical soothsayer of apocalyptic panic. Things were not endangered, extinct, or dying—they were killed, murdered. Diversity talk has substituted counting logics for engaged living. Nature is not an *object*, outside of us. We are all of nature; we need relational epistemologies. We are anti-biodiversity.

ANTI-BINARY

Edenic science and its garden of good and evil produced a settler science grounded in binary logics: life/nonlife; human/nonhuman; male/female; white/black; straight/gay; rich/poor; west/east. Even when evidence piled up to the contrary, Edenic science clung to its vanquished categories. We have done away with this. We are anti-binary. We *can* count past two.

ANTI-DISCIPLINES

Edenic science in the Edenic university vivisected a teeming, entangled, and lively world into sterile disciplines contained in claustrophobic silos policed by border walls. No wonder Edenic knowledge, with its disciplinary silos, could never solve big problems and nearly decimated the planet. Disciplines have no place in our world. We are anti-disciplines and practice anti-disciplinarity.

ANTI-CIPATORY FUTURES

In building an antiracist, anticolonial, anticasteist, antipatriarchal, antiableist world, we need to not only resist but reformulate the very ground we live on—literally and figuratively. It is through reparations, repair, and active resistance that we queered our imagination to develop antidotes to our Edenic pasts to imagine more anti-cipatory futures.

Abolitionist Futures: A Manifesto for Scientists

PREAMBLE

We are committed scientists from South Asia, a diverse group who came to the world of science for varied reasons. Inspired by movements in the United States for abolitionist education, we embrace their deep and capacious model of social justice. Some of us were drawn to science because we wanted to understand the world around us, and science promised us the necessary tools. Others wanted to make the world a better place, and science promised us powerful structures to enable this. Still others joined for the promise of an equitable world where the identity of scientist was irrelevant. Science was touted as objective and value free. Even those from socially marginalized groups could participate in it equally and fairly. Yet others came because science promised sustained employment that could help support us and our loved ones. After years of service to science and deliberations among ourselves, we are all disappointed. The institutions of science have not lived up to the promise of its possibilities.

We join with global scientists who are working toward abolitionist futures—working to undo the histories and ongoing practices of colonialism, conquest, and slavery. This much is clear to us: We recognize that there are many colonialisms, and that settler colonialism is well and alive in many parts of the world. Western colonial nations continue to control and shape the former colonies through neocolonial economic and political policies. Caste oppression continues in South Asia and its diaspora.

Science is deeply implicated in the histories of colonialism. Colonialism ushered in an extractive project that plundered colonies' resources to enrich colonial nations. This continues on to this day, where rich continue to exploit the resources of the poor. Imperialism, historians remind us, was in fact "green imperialism," an ecological project to loot nature's resources from around the globe. But most importantly, colonialism was fundamentally a knowledge project—a project that appropriated local knowledges from many peoples and regions under the mantle of western science, ultimately erasing indigenous knowledges, sciences, wisdoms, theories, practices, and epistemologies.

While independence and social movements contested colonial rule, it is clear that colonial administrators claimed "enlightenment science" as their own and instituted western science as the sole system of legitimate and universal knowledge. Most tragically, science became an instrument to rationalize the violence of colonialism. In South Asia, the British decimated social and political systems and recruited South Asians to help them run the nation. The biological sciences became critical knowledge machines to create hierarchies of gender, caste, class, race, and sexuality—the elite colonists, of course, were placed squarely at the top of this hierarchy. The tragedy is that such orientalism pathologized colonized people as inferior beings; even more tragically, this "knowledge" was institutionalized through science into governance structures and internalized within the psyche of the colonies. After independence, India was firmly gripped by a western-educated elite who, in the name of modernity and progress, imagined a nation in a western gaze. Science and technology emerged as the "reason of state," and industrialization and scientific planning became the grounding logics for the nation. Education systems and structures retain their colonial architecture to this day, and groups that inherited colonial power continue to dominate science. In India, for example, these groups are largely upper caste, Hindu, and men. In the United States, as more secure tenure track positions have become scarce, faculty of color find themselves in contingent positions with high teaching loads, their intellectual work unrecognized and their care work exploited.[1]

So deeply implicated are we in the logics of science that we have embraced it as our savior. Of those who make it into the hallways of science, many face enduring discrimination, exploitation, humiliation, and marginalization. For complex reasons, some of us continue to labor under difficult circumstances. Others are refused and excluded, and we leave. This exclusion ironically becomes transformed into a story about our incapacity, our lack of preparedness and interest. Our absence is noted as a statistic. So confident is science in its enterprise that it bemoans its exclusions and develops programs—for women, minorities, queer, and third-world scientists. At the same time, the overproduction of scientists keeps wages low. As scientists working toward liberatory ends and abolitionist sciences, we have to ask ourselves: Why do we keep sending our loved ones into the belly of the beast? Why do we not insist that the problem is science, not us? Only when science changes—when it can acknowledge its histories, when it can include the diversity that is us, when it can welcome, encourage, and incorporate our diverse viewpoints—can we be happy members of that community.

As practitioners of science living in the afterlives of empire, we see empire everywhere. For example, archives and the rich resources of the colonies were shipped off to museums and repositories in western nations for analysis and research. Living in India, we cannot access our own history. "Parachute" science lives on. Our educational system continues to be grounded in the knowledge of the west. We are so ignorant of our own history that when reactionary forces of Hindutva invent new and false histories, many of us are quick to embrace them in the name of anticolonialism. The past is not a figment of our imagination, to make and remake willy-nilly. Nor can the past be a salve for our wounded civilization. The past is a repository of our strengths and wounds. We must always be guided by rigorous method and methodologies and reckon squarely with our past. It is critical that we do this if we hope to understand where we come from and help us imagine abolitionist futures.

To this end, we believe the following:

- Everyone can understand science. Everyone is capable of doing science. We must create the educational opportunities and possibilities for inclusive institutions.
- We need to better understand and teach what science is and how science is practiced. Recent disinformation campaigns on WhatsApp University promoting cures through drinking cow urine, false stories about India's past, and hate campaigns against minorities—even among the so-called educated—have soared.
- Science is not a set of unbiased methods that produce TRUTH about the world. Rather, it is a set of historically derived knowledge practices that help elucidate the workings of the world. Science is a process; it is constantly made and remade.
- Colonialism has shaped our epistemes, methodologies, and methods. It

shapes the questions we ask, the methods we deploy, and the conclusions we derive. We must historicize science and understand where our tools come from. Only through decolonization can we ask different questions and produce new knowledge. If not, we remain bound to a colonial script.
- Colonial science was central to the shaping of human social differences—the hierarches of gender, race, class, caste, sexuality, and nation. We must always challenge the naturalization and biologization of human differences—claims of biological differences related to sex, gender, caste, race, class, sexuality, ability, or nation.
- All students of science must be taught its history. They must understand that science is produced by scientists who bring with them particular social, cultural, and political worldviews. These views shape the science they do.
- Science is not democratic. It is largely populated by members of socially dominant groups—men, upper castes, heterosexuals, adherents to dominant religions, and the economically well-off. Nor surprisingly, scientific knowledge continues to promote these very groups.
- Despite its claims, most of science is no longer basic research. Science has become the handmaiden of powerful economic forces and has increasingly moved toward consumerist, nationalist, and other powerful forces. We must reclaim a science for the people.

As practitioners of abolitionist science, we will strive to act as follows:

- Work toward a democratic and inclusive world.
- Question the status quo each day—what we have been taught, our ideas of rigor, our theories of good methods and methodologies.
- Create equitable practices in our laboratories and scientific practice.
- Recognize that a diverse science benefits all, including science.
- Recognize that everyone should have access to science and the right to question and critique science.
- Work against all forms of bias, speaking up against sexism, casteism, racism, ableism, and other forms of discrimination.
- Recognize that subjugated knowledges are valuable and that we must disagree and debate with respect.

- Rethink science and take ourselves out of the orbit of scientism. Not every problem is one that science can or must solve. We need to promote all approaches to knowledge and interdisciplinary approaches to questions. Science must not usurp all academic spaces and knowledge practices.
- Develop a science to aid people and the planet, not its wealthy funding agents and their profit margins. Science must solve everyday problems for all, not just manufacture products for the rich to buy, and must not hijack progressive concepts like sustainability into neoliberal profit motives.
- Educate ourselves about knowledges marginalized by elite science.
- Work to include excluded groups and rejected knowledges and not to reproduce colonialism by reappropriating or stealing knowledge again.
- Approach our work ethically.
- Not reduce the organisms we work with into dispensable objects but learn to work with them as worthy subjects and coinhabitants of this planet.
- Work to expand the scientific method to include indigenous and other knowledges.
- Work to recognize how systems of domination shape scientific knowledge and scientific practices.

As we work with global movements, we come together as scientists of South Asia, beyond the nations and national boundaries that are themselves colonial constructs. Decolonizing means refusing colonial maps and critically examining inherited wisdom. We celebrate our differences in our unity. We believe the practices of colonization are too deep, too entangled in our world. We must excavate our histories to understand why we inhabit a deeply unequal, violent, and inhumane world. Science has been mobilized into the necropolitics of the world—it can save lives, but in fact it saves predominantly elite ones; it produces weapons of destruction increasingly aimed at the poor and defenseless; it produces gadgets and goods that help little, even while many pressing problems loom.

We live in a world where technology increasingly mediates the world. Campaigns of misinformation and disinformation proliferate, dubbed ironically as Whatsapp University. The world is numbed by soporific social media, and people grow increasingly isolated and individualized instead of seeking friendship and community.

We remind ourselves that science is done by people. Scientific tools are tools of society. They can be used toward other ends. The world we have inherited need not be the world we live in or the world we leave to the next generation. We must work toward different futures every single day. Ends do not justify the means—decolonizing is a practice of the everyday, of every experiment, every result, every paper, and every person we meet.

We commit to working toward an abolitionist science that unravels the histories of violence, of colonialism, slavery, and conquest. We pledge to work toward more equitable and peaceful futures. This is the promise of science that brought us here. We bent to the norms of colonial science, but now we must rise to claim other futures. We are scientists; we claimed science, and now we must reclaim it.

ACKNOWLEDGMENTS

WITH AGE AND EXPERIENCE, DEBTS ACCRUE. Over the years, many individuals have profoundly shaped my life and my thinking. I owe innumerable debts for mentorship, friendship, collaborations, and collegial exchanges. Gratitude for the many feisty debates, interdisciplinary banter, and most crucially honest, frank, and brutal feedback. I have also been blessed by numerous acts of pure and unadulterated generosity in my life and in the making of this book. A list of acknowledgements can never capture how much networks of community and camaraderie ignite, nurture, and sustain a world of ideas.

I should start with plant worlds that continue to fascinate and intrigue me with their spectacular innovations and ebullience. Larin McLaughlin, my editor, and Rebecca Herzig, my series coeditor, have been a dream team to work with. Joyful, fun, astute, frank, intelligent, and insightful, their feedback has profoundly shaped this book (not only in what is in here, but in what is not!). Rebecca's brilliance as a collaborator, interlocutor, and friend has thoroughly shaped my thinking and work over the years. My deep gratitude to them both. The University of Washington Press and its staff have been a pleasure to work with. Immense gratitude to Jennifer Comeau, whose copyediting brought clarity and precision to this work. I am so grateful for her taking this on despite a busy schedule.

Having uninterrupted time to write is critically important The Samuel F. Conti Faculty fellowship allowed a productive year to complete this manuscript. Many thanks to Mike Malone for his support. I am lucky to be part of a field where senior faculty have been singularly generous, intellectually and professionally. Profound thanks to Mary Wyer, Evelynn Hammonds, Angela Ginorio, Sandra Harding, Anne Fausto Sterling, Sharon Traweek, Janet Jakobsen, Jean O'Barr, Geeta Patel, Karen Barad, Kamala Viswesaran, Ruth Hubbard, Helen Longino, Donna Haraway, Cathy Middlecamp, Miranda Joseph, Afsaneh Najmabadi, Kath Weston, Barbara Whitten, Bonnie Spanier, Evelyn Fox Keller, and Nancy Tuana. Indu Ravi has been a singularly inspirational figure. My feminist and antiestablishment spirit was kindled by her rebellious spirit during our childhood, and she remains a powerful influence in my life. Ramesh is a brilliant, curious, and cheerful presence that keeps my life engaging, unpredictable, amusing, and always joyful. Many conversations and feisty debates during long car rides have helped shape key metaphors, tropes, ideas, titles, and frameworks in this book.

Interdisciplinary work necessitates generous colleagues and friends from across the disciplines. I have been blessed with so many. I have shamelessly called on friends, colleagues, acquaintances, and well-wishers for feedback. Many of them read the full manuscript and offered insightful and helpful comments. They expanded the focus of the book and helped focus its ideas by suggesting new references, directions of thought, and theories to explore. Their feedback quite simply transformed this manuscript. Some, especially the historians on the list, were invaluable in introducing more precision in the language and temporal logics in the manuscript. My immense gratitude to colleagues: Madelaine Bartlett, Heron Breen, Karen Cardozo, Xan Chacko, Sushmita Chatterjee, Rachel Gross, Colin Hoag, Caitlin Kossmann, Ksenia Krasileva, Christa Kuljian, Greta LaFleur, Karen Lederer, Anin Luo, Michele Murphy, Samantha Pinto, Lukas Rieppel, Tara Suri, and Renny Thomas. All of them are terribly brilliant and busy, so my deepest gratitude. I am also grateful to the two anonymous reviewers for their invaluable feedback.

This widely interdisciplinary book stretched my expertise often. As a result, I called on many kind and busy colleagues whose expertise was crucial for my arguments. I drew on their careful feedback to vet particular chapters and add precision to the arguments. Their feedback helped reshape, indeed transform, the arguments: Elora Chowdhury, Michael Dietrich, Ben Goulet, Vinita Gowda, Kathleen Gutierrez, Minakshi Menon, Durba Mitra, Naima Ahmed Nash, Don Opitz, Sarah Pinto, Jyoti Puri, Beans Velocci, Kalindi Vora, and Angie Willey.

Over the years, many wonderful collaborators have proved to be lively interlocutors, shaping my thinking for the core arguments of this book. They have made academic life into a space of joy and play. Much gratitude to Sushmita Chatterjee, Rebecca Herzig, Angie Willey, Jennifer Hamilton, Deboleena Roy, Madelaine Bartlett, Mary Wyer, Betsy Hartmann, Anita Simha, Jim Bever, Peggy Schutz, Lisa Weasel, Maralee Mayberry, and Charles Zerner.

The ideas in this book were shaped by lively conversations in the book discussion group on the "vegetal turn," especially Sushmita Chatterjee, Xan Chacko, Colin Hoag, Laura Foster, Brian Sabel, and Chloe Drummond. Conversations with many colleagues and friends over the years have helped me home in on the key arguments and scope of the book. Gratitude to Itty Abraham, Neej Ahuja, Sandy Alexandre, Joseph Alter, Irina Aristarkhova, Kiran Asher, Neda Atanasoski, Aimee Bahng, Moya Bailey, Jenny Bangham, Thiago Pinto Barbosa, Rajani Bhatia, Janet Browne, Mel Chen, Nancy Chen, Giovanna di Chiro, Kathy Davis, Rob Dorit, Virginia Eubanks, Vince Formica, Nayanika Ghosh, Sig Giordano, Suzanne Gottschang, Rachel Gross, Jennifer Hamilton, Shireen Hamza, Stefan Helmreich, Nick Hopwood, Ayesha Irani, Janice Irvine, Nassim JafariNaimi, Rebecca Jordan-Young, Ambika Kamath, Katrina Karkazis, Terence Keel, Sara Khan, Michi Knecht, Yingchen Kwok, Rachel Lee, Frosty Levy, Max Liboiron, Laura Lovett, Carole McCann, Noémie Merleau-Ponty, Erika Milam, Steffan Müller-Wille, Projit Mukharji, Natasha Myers, Jennifer Nash, Kathleen Pierce, Samantha Pinto, Victoria Pitts-Taylor, Anne Pollock, Amit Prasad, Chanda Prescod-Weinstein, Anjali Prabhu, Joanna Radin, Jenny Reardon, Maria Rebolleda-Gómez, Sarah Richardson, Sophia Roosth, Deboleena Roy, Britt Rusert, Jade Sasser, Sigrid Schmitz, Katharina Schramm, Marianne Sommer, Gabriela Soto Laveaga, Chikako Takeshita, Sari van Anders, Maaike van der Lugt, Emily Varto, Harlan Weaver, Ashton Wesner, Elizabeth Wilson, and Cleo Wölfle Hazard.

UMass Amherst has been a wonderful place to work. Immense thanks to my departmental colleagues for a thoughtful and engaged working environment: Arlene Avakian, Angie Willey, Laura Briggs, Miliann Kang, Linda Hillenbrand, Karen Lederer, Alex Deschamps, Svati Shah, Kiran Asher, Kirsten Leng, Cameron Awkward-Rich, Laura Ciolkowski, Dayo Gore, and Ann Ferguson. I have benefited from stimulating conversations with many other colleagues at UMass, including Britt Rusert, Laura Doyle, Asha Nadkarni, Stephen Clingman, Susan Shapiro, Sigrid Schmalzer, Marta Calas, Brian Ogilvie, Eve Weinbaum, Jordy Rosenberg, Whitney Battle Baptiste, Sangeeta Kamat, and Fred Schaffer. Thanks also to colleagues in the Five Colleges: Jacquelyne Luce, Betsy Hartmann, Marlene Gerber Fried, Lynn Morgan, Amrita Basu, Lisa Armstrong, and Pinky Hota. During the production of this book, I moved to Wellesley College. Thanks to Kristina Jones, Elena Creef, Jenny Musto,

Betty Tiro, Jenn Yang, Emily Harrison, Adam van Arsdale, Smitha Radhakrishnan, Claire Fontijn, Tracy Tien, Lauren Savit, and Jenny O'Donnell.

Friends have sustained me in innumerable ways over the years. My graduate school buddies Becky Dunn, Jim Bever, Peggy Schultz, and Mary Malik and my childhood friends Anu Mittal and Hem Chordia Samdaria continue to ground me. My friend and neighbor Kel Moorefield has provided much joy and support over the years. Friends and family continue to make life rich and infinitely fun: Indu and J. Ravi, Martha Ayres and Arlene Avakian, Sadanand Warrier, Anuradha Warrier, Karen Lederer, Brian Sabel, Asha Ramesh, Ramesh Ratnam, Shreyas and Ani Ravi, Aditi Subramaniam, Leela and Shyam Venugopal, Shampa Chanda, V. S. Mani, Kamakshi Murti, Rosemary Kalapurakal, Vandana Date, Devamonie Naidoo, Lakshmi Ramanthan, Xan Chacko, Joshua Kim, Sepi and Dori Hashemi, Parisa Saranj, Judy Weiss, Anita Menon, Mathew Alapatt, Rudy and Taarini Subramaniam, Preeti Ravi, Varun and Vinay Sridhar, Nithya Rathinam, Ruhi Varun Nithya, P. K. Mahesh, Jayashree, and Shari and Jaan. Many thanks to Gowri and Sridhar, Rani and Prasad, Vasanthi and Joy, Srinath and Uthara Narayan, and Kamala and V. Hariharan. My mother and her sisters have been a constant inspiration, supporting my feminist ambitions in ways that have been singularly gratifying.

Earlier versions and sections of work in this book have previously appeared in journals and edited volumes: "Like a Tumbleweed in Eden: The Diasporic Life of Concepts," *Contributions to the History of Concepts* 14, no. 1 (Summer 2019): 1–16; "Affective Ecologies and the Botanical Sublime," in *Silences, Neglected Feelings, and Blind-Spots in Research Practices*, edited by Kathy Davis and Janice Irvine (New York: Routledge, 2022); "Gender and the Coloniality of Science," in "Women in Science: A Symposium on Changing Dynamics in Equity, Inclusion, Institutions, and Innovation," issue of *Seminar*, no. 760 (December 2022); "Methodologies of the Pressed: Cartographies for Adisciplinary Sciences," *Science, Technology and Society* 28, no. 1 (2022). My thanks to the editors who helped strengthen these essays.

My eternal gratitude for all who have touched my life and work. Your support has meant more than these words convey.

NOTES

Prologue

1. Betsy Hartmann, *The America Syndrome: Apocalypse, War, and Our Call to Greatness* (New York: Seven Stories Press, 2017).

2. Alfred W. Crosby, *Ecological Imperialism: The Biological Expansion of Europe, 900–1900* (New York: Cambridge University Press, 1986).

3. Charles C. Mann, *1493: Uncovering the New World Columbus Created* (New York: Vintage, 2011), 10.

4. Ocean Vyong, "Alexandra Barlyski Interview Ocean Vyong," *Marginalia*, December 21, 2018, https://themarginaliareview.com/ocean-vuong-poetry-bodies-and-stillness/ (accessed August 16, 2022); M. J. Fuentes, *Dispossessed Lives: Enslaved Women, Violence, and the Archive* (Philadelphia: University of Pennsylvania Press, 2016).

5. Eli Clare, *Brilliant Imperfection: Grappling with Cure* (Durham, NC: Duke University Press, 2017), 158.

6. Thanks to Karen Cardozo for pointing me to the multiple registers of telling history.

Introduction

1. Yota Batsaki, Sarah Burke Cahalan, and Anatole Tchikine, eds., *The Botany of Empire in the Long Eighteenth Century* (Washington, DC: Dumbarton Oaks Research Library and Collection, 2016).

2. Londa Schiebinger, *Nature's Body: Gender in the Making of Modern Science* (New Brunswick, NJ: Rutgers University Press, 2004).

3. Sam George, "Linnaeus in Letters and the Cultivation of the Female Mind: 'Botany in an English Dress,'" *Journal for Eighteenth-Century Studies* 28, no. 1 (March 2005): 1–18; quote is from 2.

4. Anna K. Sagal, *Botanical Entanglements: Women, Natural Science, and the Arts in Eighteenth-Century England* (Charlottesville: University of Virginia Press, 2022).

5. Sagal, *Botanical Entanglements*, 17.

6. Ann B. Shteir, "Gender and 'Modern' Botany in Victorian England," *Osiris* 12, no. 1 (January 1997), *Women, Gender, and Science: New Directions*: 29–38.

7. Anne Secord, "Science in the Pub: Artisan Botanists in Early Nineteenth-Century Lancashire," *History of Science* 32, no. 3 (September 1994): 269–315.

8. Anita Simha, Carlos J. Pardo-De la Hoz, and Lauren N. Carley, "Moving beyond the 'Diversity Paradox': The Limitations of Competition-Based Frameworks in Understanding Species Diversity," *American Naturalist* 200, no. 1 (July 2022).

9. Harold J. Cook, *Matters of Exchange: Commerce, Medicine, and Science in the Dutch*

Golden Age (New Haven, CT: Yale University Press, 2007); Deepak Kumar, "Botanical Explorations and the East India Company: Revisiting 'Plant Colonialism,'" in *The East India Company and the Natural World*, edited by Vinita Damodaran, Anna Winterbottom, and Alan Lester (London: Palgrave Macmillan, 2015), 16–34.

10. Richard Conniff, "A Vast Garden of Knowledge, Still Blooming Today," *New York Times*, May 13, 2007.

11. Conniff, "Vast Garden."

12. Martha Beck, "The Labyrinth of Life," *Martha's Blog*, March 2013, https://marthabeck.com/2013/03/the-labyrinth-of-life/.

13. Thanks to Karen Cardozo for this insight.

14. Audre Lorde, *Sister Outsider: Essays and Speeches* (Berkeley, CA: Crossing Press: 2012).

15. Elaine Gan, "Timing Rice: An Inquiry into More-than-Human Temporalities of the Anthropocene," *New Formations*, no. 92 (2017): 87–101.

16. John C. Ryan, "Cultural Botany: Toward a Model of Transdisciplinary, Embodied, and Poetic Research into Plants," *Nature and Culture* 6, no. 2 (Summer 2011): 128–130.

17. Vine Deloria Jr., "Ethnoscience and Indian Realities," *Winds of Change* 7, no. 3 (1992): 12–18.

18. These transformations are not possible unless we also rethink academic institutions, structures, and norms, including intellectual and personnel politics.

19. Donna Haraway, *How Like a Leaf: An Interview with Thyrza Nichols Goodve* (New York: Routledge, 1999). For a more extended discussion on naturecultures, see Banu Subramaniam, *Ghost Stories for Darwin: The Science of Variation and the Politics of Diversity* (Urbana: University of Illinois Press, 2014).

20. Patrick Wolfe, "Settler Colonialism and the Elimination of the Native," *Journal of Genocide Research* 8, no. 4 (2006): 387–409.

21. Edouard Glissant, *Caribbean Discourse: Selected Essays* (Charlottesville: University Press of Virginia, 1992), 2.

22. Aníbal Quijano, "Coloniality of Power and Eurocentrism in Latin America," *International Sociology* 15, no. 2 (June 2000): 215–232.

23. María Lugones, "Heterosexualism and the Colonial/Modern Gender System," *Hypatia* 22, no. 1 (Winter 2007): 186–219.

24. Robin Wall Kimmerer, "The Fortress, the River and the Garden: A New Metaphor for Cultivating Mutualistic Relationship between Scientific and Traditional Ecological Knowledge," in *Contemporary Studies in Environmental and Indigenous Pedagogies: A Curricula of Stories and Place*, edited by Andrejs Kulnieks, Dan Roronhiakewen Longboat, and Kelly Young (Leiden, Netherlands: Brill, 2013), 49–76.

25. Kimmerer, "The Fortress."

26. Michael-Shawn Fletcher, Rebecca Hamilton, Wolfram Dressler, and Lisa Palmer, "Indigenous Knowledge and the Shackles of Wilderness," *Proceedings of the National Academy of Sciences* 118, no. 40 (September 27, 2021): p.e2022218118: Julia E. Fa, James E. M. Watson, Ian Leiper, Peter Potapov, Tom D. Evans, Neil D. Burgess, Zsolt Molnár, et al., "Importance of Indigenous Peoples' Lands for the Conservation of Intact

Forest Landscapes," *Frontiers in Ecology and the Environment* 18, no. 3 (April 2020): 135–140; Stephen T. Garnett, Neil D. Burgess, Julia E. Fa, Álvaro Fernández-Llamazares, Zsolt Molnár, Cathy J. Robinson, James E. M. Watson, et al., "A Spatial Overview of the Global Importance of Indigenous Lands for Conservation," *Nature Sustainability* 1, no. 7 (2018): 369–374; Víctor M. Toledo, "Indigenous Peoples and Biodiversity," in *Encyclopedia of Biodiversity*, edited by Simon A. Levin, vol. 3 (San Diego: Academic Press, 2001), 451–463.

27. Vine Deloria Jr., *Red Earth, White Lies: Native Americans and the Myth of Scientific Fact* (Golden, CO: Fulcrum Publishing, 1997).

28. Enrique Salmon, "Decolonizing Our Voices," *Winds of Change* 11, no. 3 (1996): 70–72.

29. Robin Wall Kimmerer, *Braiding Sweetgrass: Indigenous Wisdom, Scientific Knowledge, and the Teachings of Plants* (Minneapolis: Milkweed Editions, 2013), and *Gathering Moss: A Natural and Cultural History of Mosses* (Corvallis: Oregon State University Press, 2021); Kriti Sharma, *Interdependence: Biology and Beyond* (New York: Fordham University Press, 2015); Cleo Wölfle Hazard, *Underflows: Queer Trans Ecologies and River Justice* (Seattle: University of Washington Press, 2022); Jessica Hernandez, *Fresh Banana Leaves: Healing Indigenous Landscapes through Indigenous Science* (Berkeley, CA: North Atlantic Books, 2022); Joan Roughgarden, *Evolution's Rainbow: Diversity, Gender, and Sexuality in Nature and People* (Los Angeles: University of California Press, 2004); Meg Lowman, *The Arbornaut: A Life Discovering the Eighth Continent in the Trees above Us* (New York: Farrar, Straus and Giroux, 2021).

30. Homi K. Bhabha, "In a Spirit of Calm Violence," in *After Colonialism: Imperial Histories and Postcolonial Displacements*, edited by Gyan Prakash (Princeton, NJ: Princeton University Press, 1994): 326–344. (284)

31. Viviane Saleh-Hanna, "Black Feminist Hauntology: Rememory the Ghosts of Abolition?" *Champ pénal/Penal Field* 12 (2015).

32. Jeong-eun Rhee, *Decolonial Feminist Research: Haunting, Rememory and Mothers* (New York: Routledge, 2020).

33. Saleh-Hanna, "Black Feminist Hauntology."

34. Christina Sharpe, *In the Wake: On Blackness and Being* (Durham, NC: Duke University Press, 2016), 22.

35. Birgit M. Kaiser and K. Thiele, "What Is Species Memory? Or, Humanism, Memory and the Afterlives of '1492,'" *Parallax* 23, no. 4 (2017): 403–415. While binaries help frame the structures of power, I am mindful that there is much more complexity. For example, colonial foot soldiers are not the same as colonial elites. Similarly, the elite of colonized nations experience the world very differently than the poor.

36. Emanuele Coccia, *The Life of Plants: A Metaphysics of Mixture* (Medford, MA: Polity Press, 2019).

37. Karen Cardozo and Banu Subramaniam, "Assembling Asian/American Naturecultures: Orientalism and Invited Invasions," *Journal of Asian American Studies* 16, no. 1 (2013).

38. Mark Sagoff, "Why Exotic Species Are Not as Bad as We Fear," *Chronicle of Higher Education* 46, no.42 (2000), B7.

39. Jennifer Terry and Jacqui L. Urla, eds., *Deviant Bodies: Critical Perspectives on Difference in Science and Popular Culture* (Bloomington: Indiana University Press, 1995).

40. For a history of eugenics on queer and disabled bodies, see Nancy Ordover, *American Eugenics: Race, Queer Anatomy, and the Science of Nationalism* (Minneapolis: University of Minnesota Press, 2003); Michael Robertson, Astrid Ley, and Edwina Light, *The First into the Dark: The Nazi Persecution of the Disabled* (Sydney, Australia: UTS ePress, 2019); Alison Bashford and Philippa Levine, eds., *The Oxford Handbook of the History of Eugenics* (New York: Oxford University Press USA, 2010); Subramaniam, *Ghost Stories*.

41. Bharat Jayram Venkat, *At the Limits of Cure* (Durham, NC: Duke University Press, 2021).

42. Discussions with Karen Cardozo have profoundly shaped my understanding of this issue.

43. Karen M. Cardozo, "Academic Labor: Who Cares?." *Critical Sociology* 43, no. 3 (2017): 405–428; Rebecca Herzig and Banu Subramaniam, "Housekeeping: Labor in the Pandemic University," *Feminist Studies* 47, no. 3 (2021): 503–517.

44. Also see Sonya Renee Taylor, *The Body Is Not an Apology: The Power of Radical Self-Love* (Berrett-Koehler Publishers, 2021).

45. Eli Clare, "Notes on Natural Worlds, Disabled Bodies, and a Politics of Cure," in *Disability and Animality: Crip Perspectives in Critical Animal Studies*, edited by Stephanie Jenkins, Kelly Struthers Montford, and Chloë Taylor (New York: Routledge, 2020); Sarah Jaquette Ray and Jay Sibara, eds., *Disability Studies and the Environmental Humanities: Toward an Eco-Crip Theory* (Lincoln: University of Nebraska Press, 2017); Leah Lakshmi Piepzna-Samarasinha, *Care Work: Dreaming Disability Justice* (Vancouver: Arsenal Pulp Press, 2018).

46. Raquel Velho, *Hacking the Underground: Disability, Infrastructure, and London's Public Transport System* (Seattle: University of Washington Press, 2023).

47. Aimi Hamraie and Kelly Fritsch, "Crip Technoscience Manifesto," *Catalyst: Feminism, Theory, Technoscience* 5, no. 1 (2019): 1–33.

48. Eli Clare, *Exile and Pride: Disability, Queerness, and Liberation* (Cambridge, MA: South End Press, 1999), 2.

49. Alison Kafer, *Feminist, Queer, Crip* (Bloomington: Indiana University Press, 2013).

50. Sumana Roy, *How I Became a Tree* (New Haven, CT: Yale University Press, 2021).

51. Petra Kuppers, "Crip Time," *Tikkun* 29, no. 4 (Fall 2014): 29–30.

52. Kim Q. Hall, *Feminist Disability Studies* (Indiana University Press, 2011).

53. For links to critical animal studies, see Stephanie Jenkins, Kelly Struthers Montford, and Chloë Taylor, eds., *Disability and Animality: Crip Perspectives in Critical Animal Studies* (New York: Routledge, 2020); Ray and Sibara, *Disability Studies and the Environmental Humanities*; Stacy Alaimo, *Bodily Natures: Science, Environment, and the Material Self* (Bloomington: Indiana University Press, 2010).

54. Michel Foucault, *The History of Sexuality: An Introduction* (Vintage, 1990).

55. Eve K. Sedgwick, *Tendencies* (New York: Routledge, 1994), 7.

56. Cyd Cipolla, Kristina Gupta, David Rubin, and Angela Willey, eds., *Queer Feminist Science Studies: A Reader* (Seattle: University of Washington Press, 2017).

57. Angela Willey, *Undoing Monogamy: The Politics of Science and the Possibilities of Biology* (Duke University Press, 2016).

58. J. Jack Halberstam, *In a Queer Time and Place: Transgender Bodies, Subcultural Lives* (New York: NYU Press, 2005).

59. I should stress that there are similar epistemic tensions and ambiguities in the space of imperial zoology and in social evolutionary theories.

60. Wölfle Hazard, *Underflows*.

61. Sunaura Taylor, "Disabled Ecologies: Living with Impaired Landscapes," *Haas Institute for a Fair and Inclusive Society*. 2019: https://www.youtube.com/watch?v=_OOEXLylhT4.

62. Many thanks to Rebecca Herzig for this wonderful phrasing.

63. Az Causevic, Kavita Philip, Maari Zwick-Maitreyi, Persephone Hooper Lewis, Siko Bouterse, and Anasuya Sengupta, "Centering Knowledge from the Margins: Our Embodied Practices of Epistemic Resistance and Revolution," *International Feminist Journal of Politics* 22, no. 1 (2020): 6–25.

64. Jorge V. Crisci, Liliana Katinas, María J. Apodaca, and Peter C. Hoch, "The End of Botany," *Trends in Plant Science* 25, no. 12 (2020): 1173–1176.

65. Allie Bidwell, "The Academic Decline: How to Train the Next Generation of Botanists," *US News & World Report*, November 12, 2013.

66. Daniel S. Park, Xiao Feng, Shinobu Akiyama, Marlina Ardiyani, Neida Avendaño, Zoltan Barina, Blandine Bärtschi, et al., "The Colonial Legacy of Herbaria," *bioRxiv* (posted November 2, 2021): https://doi.org/10.1101/2021.10.27.466174.

67. Schiebinger, *Nature's Body*.

68. Schiebinger, *Nature's Body*, 12.

69. While *repatriation* signals a legal return of stolen land, *rematriation* has emerged as a more profound and decolonizing term that calls for restoring indigenous relations between land and communities.

70. Minakshi Menon, "Gardens of Empire: On the Politics of Collecting Nature," *Humboldt Forum*, 2002c: https://www.youtube.com/watch?v=2mm0A9uZ8zQ&t=6815s.

71. Samantha Klein, James Sanghyun Lee, Sofi Courtney, Lisa Morehead-Hillman, Sallie Lau, Bryce Lewis-Smith, Daniel Sarna-Wojcicki, and Cleo Woelfle-Hazard, "Transforming Restoration Science: Multiple Knowledges and Community Research Cogeneration in the Klamath and Duwamish Rivers," *American Naturalist* 200, no. 1 (July 2022).

72. Londa Schiebinger and Claudia Swan, eds., *Colonial Botany: Science, Commerce, and Politics in the Early Modern World* (Philadelphia: University of Pennsylvania Press, 2007).

73. John Law and Wen-yuan Lin, "Provincializing STS: Postcoloniality, Symmetry, and Method," *East Asian Science, Technology and Society: An International Journal* 11, no. 2 (2017): 211–227.

74. Many thanks to Karen Cardozo for her work and our many discussions. See Davarian L. Baldwin, *In the Shadow of the Ivory Tower: How Universities Are Plundering Our Cities* (Bold Type Books, 2021); Sara Ahmed, *On Being Included: Racism and Diversity in Institutional Life* (Durham, NC: Duke University Press, 2020).

75. Gideon F. Smith, Estrela Figueiredo, Timothy A. Hammer, and Kevin R. Thiele, "Dealing with Inappropriate Honorifics in a Structured and Defensible Way Is Possible," *Taxon* 71, no. 5 (2022): 933–935; Kevin R. Thiele, Gideon F. Smith, Estrela Figueiredo, and Timothy A. Hammer, "Taxonomists Have an Opportunity to Rid Botanical Nomenclature of Inappropriate Honorifics in a Structured and Defensible Way," *Taxon* 71, no. 5 (2022): 1151–1154.

76. Michael Pollan, *The Omnivore's Dilemma: A Natural History of Four Meals* (Penguin, 2007); Ursula Lang, *Living with Yards: Negotiating Nature and the Habits of Home* (McGill-Queen's Press-MQUP, 2022).

CHAPTER ONE | *The Botanical Sublime*

1. Richard H. Grove, *Green Imperialism: Colonial Expansion, Tropical Island Edens and the Origins of Environmentalism, 1600–1860* (Cambridge: Cambridge University Press, 1995); Crosby, *Ecological Imperialism*.

2. Sidney W. Mintz, *Sweetness and Power: The Place of Sugar in Modern History* (New York: Penguin Books 1985); Sven Beckert, *Empire of Cotton: A Global History* (New York: Vintage Books, 2014).

3. Grove, *Green Imperialism*; William Beinart and Lotte Hughes, *Environment and Empire* (Oxford: Oxford University Press, 2007); Graham Huggan and Helen Tiffin, *Postcolonial Ecocriticism: Literature, Animals, Environment* (New York: Routledge, 2015).

4. As Ann Shteir demonstrates, women were systematically pushed out of the study of plants as botany became a masculine "science." Ann Shteir, *Cultivating Women, Cultivating Science: Flora's Daughters and Botany in England, 1760–1860* (Baltimore: Johns Hopkins University Press, 1999).

5. I have sketched out a few examples here. For more detail and other examples see Nandita Krishna and M. Amirthalingam, *Sacred Plants of India* (India Penguin, 2014).

6. For a greater discussion of Indian mythological storytelling tradition and science, see Banu Subramaniam, *Holy Science: The Biopolitics of Hindu Nationalism* (Seattle: University of Washington Press, 2019).

7. Thanks to Colin Hoag for this wonderful insight.

8. Kimmerer, *Braiding Sweetgrass*, 41–42.

9. Jamaica Kincaid, *Lucy: A Novel* (New York: Macmillan, 1990).

10. Patrick Barkham, "The Real David Attenborough," *The Guardian*, October 22, 2019: https://www.theguardian.com/tv-and-radio/2019/oct/22/david-attenborough-climate-change-bbc.

11. Andrew Anthony, "Is There Life on Earth after Attenborough?" *The Guardian*, June 12, 2011: https://www.theguardian.com/environment/2011/jun/12/life-on-earth-after-david-attenborough.

12. Also see Jack Halberstam, *Wild Things: The Disorder of Desire* (Durham, NC: Duke University Press, 2020); and Tavia Nyong'o, "Little Monsters: Race, Sovereignty, and Queer Inhumanism in *Beasts of the Southern Wild*," *GLQ: A Journal of Lesbian and Gay Studies* 21, no. 2–3 (2015): 249–272.

13. For a more detailed discussion see Subramaniam, *Ghost Stories*.

14. Banu Subramaniam and Mary Wyer, "Assimilating the 'Culture of No Culture' in Science: Feminist Interventions in (De)Mentoring Graduate Women," *Feminist Teacher* 12, no. 1 (1998): 12–28.

15. David F. Noble, *A World without Women: The Christian Clerical Culture of Western Science* (Oxford: Oxford University Press, 1992).

16. David Wade Chambers, "Stereotypic Images of the Scientists: The Draw-a-Scientist Test," *Science Education* 67, no. 2 (1983): 255–265. (256).

17. Noble, *A World without Women*.

18. While working on this project, I discovered that some of my peers in graduate school who grew up in urban areas in the U.S, also felt a similar alienation. It is significant that we did not share this feeling while in graduate school.

19. I should note that unlike me, and some urban American counterparts, many Indians grow up around fields and mountains (and not just the privileged). Yet they are less likely to find themselves in a graduate program in biology in the United States.

20. Meera Nanda, *The God Market: How Globalization Is Making India More Hindu* (New York: NYU Press, 2011).

21. Carolyn Finney, *Black Faces, White Spaces: Reimagining the Relationship of African Americans to the Great Outdoors* (Chapel Hill: University of North Carolina Press, 2014); Camille T. Dungy, *Black Nature: Four Centuries of African American Nature Poetry* (Athens: University of Georgia Press, 2009).

22. Sharon Traweek, *Beamtimes and Lifetimes: The World of High Energy Physicists* (Cambridge, MA: Harvard University Press, 1988).

23. Banu Subramaniam, Rebecca Dunn, and Lynn Broaddus, "'Sir'vey or 'Her'vey," in *Engaging Feminism: Students Speak Up and Speak Out,* edited by Jean O'Barr and Mary Wyer (Chicago: University of Chicago Press, 1992).

24. I owe an immense depth of gratitude to my graduate school friends and collaborators in this work, especially Rebecca Dunn, Peggy Schultz, Jim Bever, and Mary Malik. The generous mentorship of Jean O'Barr and Mary Wyer was critical to all of us.

25. Traweek, *Beamtimes and Lifetimes*.

26. Lesley A. Sharp, *Animal Ethos: The Morality of Human-Animal Encounters in Experimental Lab Science* (Berkeley: University of California Press, 2018).

27. Subramaniam and Wyer, "Assimilating the 'Culture of No Culture.'"

28. Evelyn Fox Keller, *Reflections on Gender and Science* (New Haven, CT: Yale University Press, 1985), 174.

29. The choice of morning glories was a practical decision. I needed an annual plant for my work, and the laboratory I worked in used morning glories as its experimental model, hence an obvious choice.

30. Evelyn Fox Keller, *A Feeling for the Organism: The Life and Work of Barbara McClintock* (San Francisco: W. H. Freeman, 1983).

31. Kimmerer, *Braiding Sweetgrass*, 252.

32. While I can only speak for my experiences, many friends and colleagues from other fields and disciplines have also claimed similar experiences.

33. Historians have chronicled many important moments in the history of science. For example, for science as adventure see Joshua A. Bell and Erin Hasinof, eds., *The Anthropology of Expeditions: Travel, Visualities, Afterlives* (Chicago: University of Chicago Press, 2015); or science as suffering see Herzig, R., *Suffering for Science: Reason and Sacrifice in Modern America* (New Brunswick, NJ: Rutgers University Press, 2005).

34. Philip Shaw, *The Sublime: The Critical Idiom* (New York: Routledge, 2006).

35. Shaw, *The Sublime*, 2

36. Theodore E. B. Wood, *The Word Sublime and Its Context, 1650–1760* (The Hague: Mouton, 1972); James Noggle, *The Skeptical Sublime: Aesthetic Ideology in Pope and the Tony Satirists* (New York: Oxford University Press, 2001); Timothy Costelloe, *The Sublime: From Antiquity to the Present* (New York: Cambridge University Press, 2012).

37. Esa Kirkkopelto, "Farewell to the Sublime? Performance Criticism in the Age of Terrorism," *Forum Modernes Theater* 29, no. 1–2 (2014): 47–55.

38. Costelloe, *The Sublime*.

39. Emily Brady, *The Sublime in Modern Philosophy: Aesthetics, Ethics, and Nature* (New York: Cambridge University Press, 2013).

40. Richard W. Stoffle, Rebecca Toupal, and Nieves Zedeño, "Landscape, Nature, and Culture: A Diachronic Model of Human-Nature Adaptations," *Nature across Cultures: Views of Nature and the Environment in Non-Western Cultures* (2003): 97–114.

41. William Cronon, "The Trouble with Wilderness: or, Getting Back to the Wrong Nature," *Environmental History* 1, no. 1 (January 1996): 9–10.

42. Cronon, "The Trouble with Wilderness," 10.

43. Cronon, "The Trouble with Wilderness," 11.

44. Cronon, "The Trouble with Wilderness," 13.

45. Leo Marx, *The Machine in the Garden: Technology and the Pastoral Ideal in America* (New York: Oxford University Press, 1964): 198.

46. Michael Marder, *The Philosopher's Plant: An Intellectual Herbarium* (New York: Columbia University Press, 2014), 1.

47. Henry David Thoreau, *Journal Entry*, April 2, 1852: https://www.walden.org/collection/journals/.

48. Carolyn Merchant, *The Death of Nature: Women, Ecology, and the Scientific Revolution* (San Francisco: Harper & Row, 1980).

49. Richard Grove, *Green Imperialism*; Crosby, *Ecological Imperialism*.

50. S. P. Gasteyer and C. Butler Flora, "Modernizing the Savage: Colonization and Perceptions of Landscape and Lifescape," *Sociologia Ruralis* 40, no. 1 (2000): 128–149.

51. Dipesh Chakrabarty, *Provincializing Europe: Postcolonial Thought and Historical Difference*, new ed. (Princeton, NJ: Princeton University Press, 2008). Many thanks to Xan Chacko for this key phrasing.

52. Cook, *Matters of Exchange*.

53. Schiebinger and Swan, *Colonial Botany*.

54. Hannah Holleman, *Dust Bowls of Empire* (New Haven, CT: Yale University Press, 2018); Philip, Kavita. "Seeds of Neo-colonialism? Reflections on Ecological Politics in the New World Order," *Capitalism Nature Socialism* 12, no. 2 (2001): 3–47; Kavita Philip, "Nature, Culture Capital, Empire," *Capitalism Nature Socialism* 18, no. 1 (2007):

5–12; Hilary Rose. *Love, Power, and Knowledge: Towards a Feminist Transformation of the Sciences* (Bloomington: Indiana University Press, 1994); Richard Levins, and Richard Lewontin, *The Dialectical Biologist* (Cambridge, MA: Harvard University Press, 1985).

55. Rajasri Ray and Madhupreeta Muralidhar, "Spatio-Temporal Patterns in the History of Colonial Botanical Exploration in India," *Endeavour* 47 (2023): 100859.

56. Lucile Brockway, *Science and Colonial Expansion: The Role of the British Royal Botanic Gardens* (New Haven, CT: Yale University Press, 2002).

57. Donna Haraway, *How Like a Leaf: An Interview with Thyrza Nichols Goodeve* (New York: Routledge, 2000).

CHAPTER TWO | *The Coloniality of Botany*

1. Florike Egmond, "The Garden of Nature: Visualizing Botanical Research in Northern and Southern Europe in the 16th Century," in *From Art to Science: Experiencing Nature in the European Garden, 1500–1700*, edited by Juliette Ferdinand (Merlengo: ZeL Edizioni, 2016), 18–33.

2. Gayatri Spivak, *A Critique of Postcolonial Reason: Toward a History of the Vanishing Present* (Cambridge, MA: Harvard University Press, 1999), 208.

3. Priya Satia, *Time's Monster: How History Makes History* (Cambridge, MA: Harvard University Press, 2020).

4. Quoted in Amitav Ghosh, "Amitav Ghosh on Priya Satia's Books: 'History Has Given Us Tools for Upending Dominant Narratives,'" *Scroll*, June 26, 2021: https://scroll.in/article/998495/amitav-ghosh-on-priya-satias-books-history-has-given-us-tools-for-upending-dominant-narratives.

5. Ghosh, "Amitav Ghosh on Priya Satia's Books."

6. Leanne Betasamosake Simpson, "Not Murdered, Not Missing: Rebelling against Colonial Gender Violence," in *Burn It Down! Feminist Manifestos for the Revolution*, edited by Breanne Fahs (London: Verso, 2020).

7. Irene Watson, "Aboriginal Relationships to the Natural World: Colonial 'Protection' of Human Rights and the Environment," *Journal of Human Rights and the Environment* 9, no. 2 (2018): 119–140.

8. Ania Loomba, *Colonialism/Postcolonialism* (New York: Routledge, 2007).

9. Colonialism is a recurrent and widespread feature of history. For example, consider the Roman Empire, Mongols under Genghis Kahn, the Aztec Empire, the Ottoman Empire, the Vijaynagar Empire, and the Chinese empire among others (Loomba, *Colonialism/Postcolonialism*). On the violence of European colonialism, see Quijano, "Coloniality of Power"; Loomba, *Colonialism/Postcolonialism*; Walter Mignolo, "Introduction: Coloniality of Power and Decolonial Thinking," *Cultural Studies* 21, no. 2–3 (2007): 155–167; and Sylvia Wynter, "Unsettling the Coloniality of Being/Power/Truth/Freedom: Towards the Human, after Man, Its Overrepresentation—An Argument," *CR: The New Centennial Review* 3, no. 3 (2003): 257–337.

10. Ramachandra Guha, "Forestry in British and Post-British India: A Historical Analysis," *Economic and Political Weekly* 18, no. 44 (1983): 1882–1896.

11. Patrick Wolfe, *Traces of History: Elementary Structures of Race* (London: Verso, 2016).

12. Patricia Hill Collins, *Black Feminist Thought: Knowledge, Consciousness, and the Politics of Empowerment*, 2nd ed. (New York: Routledge, 1999).

13. Harriet A. Washington, *Medical Apartheid: The Dark History of Medical Experimentation on Black Americans from Colonial Times to the Present* (New York: Doubleday, 2006).

14. Kiran Asher, "Latin American Decolonial Thought, or Making the Subaltern Speak," *Geography Compass* 7, no. 12 (2013): 832–842.

15. Quijano, "Coloniality of Power."

16. Nandita Sharma, "Strategic Anti-Essentialism: Decolonizing Decolonization," in *Sylvia Wynter: On Being Human as Praxis*, ed. Katherine McKittrick (Durham, NC: Duke University Press, 2015): 164–182.

17. Audra Simpson, *Theorizing Native Studies* (Durham, NC: Duke University Press, 2014).

18. Eve Tuck and K. Wayne Yang, "Decolonization Is Not a Metaphor," *Decolonization: Indigeneity, Education & Society* 1, no.1 (2012): 1–40.

19. Fanon, *The Wretched of the Earth.*

20. Kapil Raj, *Relocating Modern Science: Circulation and the Construction of Knowledge in South Asia and Europe, 1650–1900* (Springer, 2007); Amit Prasad, "Burdens of the Scientific Revolution: Euro/West-Centrism, Black Boxed Machines, and the (Post) Colonial Present," *Technology and Culture* 60, no. 4 (2019): 1059–1082.

21. S. Srivastav and R. Balabsubramanian, "Sustainability—A Decolonizing Practice," *India-Seminar* 750 (February 2022).

22. Rolando Vazquez, "Precedence, Earth and the Anthropocene: Decolonizing Design," *Design Philosophy Papers* 15, no. 1 (2017): 77–91.

23. Eli Nelson, "Knowing and Gendering the NDN Cyborg," *Catalyst: Feminism, Theory, Technoscience* 7, no. 1 (2017): 1–34.

24. Max Liboiron, *Pollution Is Colonialism* (Durham, NC: Duke University Press, 2021), 42.

25. Kuokkanen quoted in Adam Gaudry and Danielle Lorenz, "Indigenization as Inclusion, Reconciliation, and Decolonization: Navigating the Different Visions for Indigenizing the Canadian Academy," *AlterNative: An International Journal of Indigenous Peoples* 14, no. 3 (2018): 218.

26. Gaudry and Lorenz, "Indigenization as Inclusion."

27. Anna M. Lawrence, "Listening to Plants: Conversations between Critical Plant Studies and Vegetal Geography," *Progress in Human Geography* 46, no. 2 (2022): 629–651.

28. Ruakere Hond, Mihi Ratima, and Will Edwards, "The Role of Māori Community Gardens in Health Promotion: A Land-Based Community Development Response by Tangata Whenua, People of Their Land," *Global Health Promotion* 2 (2019): 44–53; Lara A. Jacobs, Coral B. Avery, Rhode Salonen, and Kathryn D. Champagne, "Unsettling Marine Conservation: Disrupting Manifest Destiny–Based Conservation Practices through the Operationalization of Indigenous Value Systems," *Parks Stewardship Forum* 38, no. 2 (2022); Laura Peach, Chantelle A. M. Richmond, and Candace Brunette-Debassige, "'You Can't Just Take a Piece of Land from the University and Build a Garden on It': Exploring Indigenizing Space and Place in a Settler Canadian University Context," *Geoforum* 114 (August 2020): 117–127.

29. Kim TallBear, "Caretaking Relations, Not American Dreaming," *Kalfou* 6, no. 1 (2019), 24–41.

30. Minakshi Menon, "Indigenous Knowledges and Colonial Sciences in South Asia," *South Asian History and Culture* 13, no. 1 (2022): 1–18.

31. Vivek Gupta, quoted in Kabir Jhala, "At a Cambridge University College, Wrestling with Its Imperial Past, Shahzia Sikander's Show Offers New Ideas on Restitution," *Art Newspaper*, December 9, 2021, https://www.theartnewspaper.com/2021/12/09/shahzia-sikander-jesus-college-cambridge-restitution-colonial-loot.

32. Kim TallBear, review of *All Our Relations: Native Struggles for Land and Life* by Winona LaDuke, *Wicazo Sa Review* 17, no. 1 (Spring 2002): 234–242.

33. Warwick Anderson, "Introduction: Postcolonial Technoscience," *Social Studies of Science* 32, no. 5–6 (2002): 643–658; Itty Abraham, "The Contradictory Spaces of Postcolonial Techno-Science," *Economic and Political Weekly* 41, no. 3 (January 21–27, 2006): 210–217; Lilly Irani, Janet Vertesi, Paul Dourish, Kavita Philip, and Rebecca E. Grinter, "Postcolonial Computing: A Lens on Design and Development," in *Proceedings of the SIGCHI conference on human factors in computing systems*, 2010: 1311–1320; Sandra Harding, *The Postcolonial Science and Technology Studies Reader* (Durham, NC: Duke University Press, 2011); Suman Seth, "Colonial History and Postcolonial Science Studies," *Radical History Review* 127 (January 1, 2017): 63–85; Banu Subramaniam, Laura Foster, Sandra Harding, Deboleena Roy, and Kim TallBear, "Feminism, Postcolonialism, Technoscience," *The Handbook of Science and Technology Studies*, ed. Ulrike Felt, Rayvon Fouché, Clark A. Miller, and Laurel Smith-Doerr (Cambridge, MA: MIT Press, 2016), 407.

34. Kristina Lyons, Juno Salazar Parreñas, Noah Tamarkin, Banu Subramaniam, Lesley Green, and Tania Pérez-Bustos, "Engagements with Decolonization and Decoloniality in and at the Interfaces of STS," *Catalyst: Feminism, Theory, Technoscience* 3, no. 1 (2017): 1–47: https://catalystjournal.org/index.php/catalyst/article/view/28794/html_11.

35. Yarimar Bonilla and Marisol LeBrón, eds., *Aftershocks of Disaster: Puerto Rico before and after the Storm* (Chicago: Haymarket Books, 2019).

36. Ryan Cecil Jobson, "The Case for Letting Anthropology Burn: Sociocultural Anthropology in 2019," *American Anthropologist* 122, no. 2 (2020): 259–271.

37. Ben Ehrenreich, *Desert Notebooks: A Road Map for the End of Time* (Berkeley, CA: Counterpoint Press, 2020).

38. Amitav Ghosh, *The Nutmeg's Curse: Parables for a Planet in Crisis* (Chicago: University of Chicago Press, 2021).

39. Ghosh, *The Nutmeg's Curse*, 55.

40. William Cronon, *Changes in the Land: Indians, Colonists, and the Ecology of New England* (New York: Hill and Wang, 2011).

41. Kimmerer, *Braiding Sweetgrass*, 6.

INTERLUDE | *Revisiting the "Women in Science" Question*

1. Margaret Rossiter, *Women Scientists in America: Struggles and Strategies to 1940* (Baltimore: Johns Hopkins University Press, 1982); Margaret Rossiter, *Women Scientists in America: Before Affirmative Action, 1940–1972* (Baltimore: Johns Hopkins University

Press, 1995); Neelam Kumar, ed., *Women and Science in India: A Reader* (New Delhi: Oxford University Press, 2009); Abigail Stewart and Virginia Valian, *Achieving Diversity and Excellence* (Cambridge, MA: MIT Press, 2018); Arjee Restar and Don Operario, "The Missing Trans Women of Science, Medicine, and Global Health," *The Lancet* 393, no. 10171 (2019): 506–508, Rita Colwell, Asheley Bear, and Alex Helman, eds., *Promising Practices for Addressing the Underrepresentation of Women in Science, Engineering, and Medicine: Opening Doors* (Washington, DC: The National Academies Press, 2020), Donald L. Opitz, "Domesticities and the Sciences," *Histories* 2, no. .3 (2022): 259–269.

2. Subramaniam, *Ghost Stories.*

3. Chandra Talpade Mohanty, *Feminism without Borders: Decolonizing Theory, Practicing Solidarity* (Durham, NC: Duke University Press, 2003); Benjamin Baez, "The Study of Diversity: The 'Knowledge of Difference' and the Limits of Science," *Journal of Higher Education* 75, no. 3 (2004): 285–306; Ahmed, *On Being Included*; Gabriella Gutiérrez y Muhs, Yolanda Flores Niemann, Carmen G. Gonzalez, and Angela P. Harris, eds., *Presumed Incompetent: The Intersections of Race and Class for Women in Academia* (Logan: Utah State University Press, 2012); Fred Moten and Stefano Harney, *The Undercommons: Fugitive Planning and Black Study* (Chico, CA: AK Press, 2013); Subramaniam, *Ghost Stories*; Nathan Snaza and Julietta Singh, "Introduction: Dehumanist Education and the Colonial University," *Social Text* 39, no. 1 (2021): 1–19.

4. Baez, "The Study of Diversity"; Ahmed, *On Being Included.*

5. Mohanty, *Feminism without Borders*, 193.

6. Mathew Boyle, "More Workers Ready to Quit over 'Window Dressing' Racism Efforts," *Bloomberg*, June 9, 2022.

7. Subramaniam, *Ghost Stories.*

8. Cinda-Sue Davis, Angela B. Ginorio, Carol S. Hollenshead, Barbara B. Lazarus, Paula M. Rayman et al., *The Equity Equation: Fostering the Advancement of Women in the Sciences, Mathematics, and Engineering* (San Francisco: Jossey-Bass, 1996).

9. Lisa Garforth and Anne Kerr, "Women and Science: What's the Problem?" *Social Politics: International Studies in Gender, State and Society* 16, no. 3 (2009): 379–403.

10. Noble, *A World without Women*, 163.

11. For a discussion of the transition to familial life, see Gadi Algazi, "Scholars in Households: Refiguring the Learned Habitus, 1480–1550," *Science in Context* 16, no. 1–2 (2003), 9–42.

12. Garforth and Kerr, "Women and Science."

13. Wynter, "Unsettling the Coloniality"; Catherine Walsh, "Shifting the Geopolitics of Critical Knowledge," *Cultural Studies* 21, no. 2–3 (2007): 224–239; Lugones, "Heterosexualism and the Colonial/Modern Gender System."

14. Donna J. Haraway, *Modest_Witness@Second_Millennium.FemaleMan_Meets_OncoMouse: Feminism and Technoscience* (New York: Routledge, 2018).

15. S. Ramaseshan, "A Conversation with Satyendranath Bose about Five Decades Ago: Some Recollections," *Current Science* 78, no. 5 (March 10, 2000).

16. Sandra Harding, *Is Science Multicultural? Postcolonialisms, Feminisms, and Epistemologies* (Bloomington: Indiana University Press, 1988).

17. Richard Grove, "Indigenous Knowledge and the Significance of South-West India

for Portuguese and Dutch Constructions of Tropical Nature," *Modern Asian Studies* 30, no. 1 (1996): 121–143; Crosby, *Ecological Imperialism*; Raj, *Relocating Modern Science*; Schiebinger and Swan, *Colonial Botany*.

18. Raj, *Relocating Modern Science*; Kavita Philip, *Civilizing Natures: Race, Resources, and Modernity in Colonial South India* (New Brunswick, NJ: Rutgers University Press, 2004); Projit Bihari Mukharji, "Vishalyakarani as Eupatorium Ayapana: Retro-Botanizing, Embedded Traditions, and Multiple Historicities of Plants in Colonial Bengal, 1890–1940," *Journal of Asian Studies* 73, no. 1 (2014): 65–87; Sumi Krishna and Gita Chadha, eds., *Feminists and Science: Critiques and Changing Perspectives in India* (SAGE Publishing India, 2017).

19. Marwa Elshakry, "When Science Became Western: Historiographical Reflections," *Isis* 101, no. 1 (2010): 98–109.

20. Rose A. Marks, Erik J. Amézquita, Sarah Percival, Alejandra Rougon-Cardoso, Claudia Chibici-Revneanu, Shandry M. Tebele, Jill M. Farrant, Daniel H. Chitwood, and Robert VanBuren, "A Critical Analysis of Plant Science Literature Reveals Ongoing Inequities," *Proceedings of the National Academy of Sciences* 120, no. 10 (2023), p.e2217564120.

21. Caitlin Karniski and Carol Ibe, "Training the Trainers: An Interview with Carol Ibe on the Importance Of Building Networks for Agricultural Research in African Countries," *Communications Biology* 4, no. 1 (2021).

22. Many thanks to Madelaine Bartlett for this insight.

23. Estrela Figueiredo and Gideon Smith, "The Colonial Legacy in African Plant Taxonomy: Biological Types and Why We Need Them," *South African Journal of Science* 106, no. 3–4 (2010): 1–4.

24. Helen Verran, "A Postcolonial Moment in Science Studies: Alternative Firing Regimes of Environmental Scientists and Aboriginal Landowners," *Social Studies of Science* 32, no. 5–6 (2002): 729–762.

25. Linda Tuhiwai Smith, *Decolonizing Methodologies* (London: Zed Books, 2010): 33.

CHAPTER THREE | *The Categorical Impurative*

1. Quoted in Peter Austin, *1000 Languages: The Worldwide History of Living and Lost Tongues* (London: Thames & Hudson, 2008).

2. Krzysztof Iwanek, "Destiny in Numbers: How Numerology Is Ruining Indian Names and Traditions," *The Diplomat*, October 4, 2017.

3. Geoffrey C. Bowker, "The Game of the Name: Nomenclatural Instability in the History of Botanical Informatics," in *Proceedings of the 1998 Conference on the History and Heritage of Science Information Systems*, ed. Mary Ellen Bowden, Trudi Bellardo Hahn, and Robert V. Williams (Medford, NJ: Published for the American Society for Information Science and the Chemical Heritage Foundation by Information Today, 1999), 74–83.

4. Peter H. Raven, Brent Berlin, and Dennis E. Breedlove, "The Origins of Taxonomy: A Review of Its Historical Development Shows Why Taxonomy Is Unable to Do What We Expect of It," *Science* 174, no. 4015 (1971): 1210–1213.

5. Brent Berlin, "Folk Systematics in Relation to Biological Classification and Nomenclature," *Annual Review of Ecology and Systematics*, 1973: 259–271. Thanks to Greta LaFleur for reminding me that this framing is similar to what Foucault calls *subjugated knowledges/anti-sciences* (Michel Foucault, *The Order of Things: An Archaeology of the Human Sciences* [New York: Vintage, 1970], 56).

6. David L. Hull, "Species, Subspecies, and Races," *Social Research* 65, no. 2 (Summer 1998): 353.

7. Jim Endersby, "Descriptive and Prescriptive Taxonomies," in *Worlds of Natural History*, ed. Helen Anne Curry et al. (New York: Cambridge University Press, 2018).

8. Ernst Mayr, "Cladistic Analysis or Cladistic Classification?" *Journal of Zoological Systematics and Evolutionary Research* 12, no. 1 (1974): 94–128.

9. Germinal Rouhan and Myriam Gaudeul, "Plant Taxonomy: A Historical Perspective, Current Challenges, and Perspectives," *Molecular Plant Taxonomy* 1115 (2014): 1–38.

10. Gavin Hardy and Laurence Totelin, *Ancient Botany* (New York: Routledge, 2015); S. K. Jain and Harsh Singh, "India's Notable Presence in Linnaeus' Botanical Classification," *Indian Journal of History of Science* 49, no. 1 (2014): 34–41.

11. Mayr, "Cladistic Analysis?"

12. David Gledhill, *The Names of Plants* (New York: Cambridge University Press, 2008).

13. Brian W. Ogilvie, "Beasts, Birds, and Insects: Folkbiology and Early Modern Classification of Insects," *Wissenschaftsgeschichte und Geschichte des Wissens im Dialog—Connecting Science and Knowledge* (Göttingen: V & R Unipress, 2013): 295–316.

14. Allen J. Grieco, "The Social Politics of Pre-Linnaean Botanical Classification," *I Tatti Studies: Essays in the Renaissance* 4 (1991): 131–49.

15. Ogilvie, "Beast, Birds, and Insects."

16. Ogilvie, "Beast, Birds, and Insects."

17. Thanks to Minakshi Menon for this insight. See Julia Graves, *The Language of Plants: A Guide to the Doctrine of Signatures* (Great Barrington, MA: Lindisfarne Books, 2012).

18. Foucault, *The Order of Things*.

19. Jean Feerick, "Botanical Shakespeares: The Racial Logic of Plant Life in Titus Andronicus," *South Central Review* 26, no. 1 (2009): 82–102.

20. Jim Endersby, *Imperial Nature: Joseph Hooker and the Practices of Victorian Science* (Chicago: University of Chicago Press, 2020).

21. Carol Kaesuk Yoon, *Naming Nature: The Clash between Instinct and Science* (New York: W. W. Norton & Company, 2009), 123.

22. Yoon, *Naming Nature*, 43.

23. Helen Pearson, "250 Years of Linnaeus' Plant Names Celebrated," *Nature* 29 (2003).

24. Endersby, *Imperial Nature*, 172.

25. Mayr, "Cladistic Analysis?"

26. Crair, "The Bizarre Bird."

27. Schiebinger and Swan, *Colonial Botany*; Laurelyn Whitt, *Science, Colonialism, and Indigenous Peoples: The Cultural Politics of Law and Knowledge* (New York: Cambridge University Press, 2009).

28. Cori Hayden, *When Nature Goes Public: The Making and Unmaking of*

Bioprospecting in Mexico (Princeton, NJ: Princeton University Press, 2003); Vandana Shiva, *Biopiracy: The Plunder of Nature and Knowledge* (Berkeley, CA: North Atlantic Books, 2016).

29. Stefano Mancuso, *The Revolutionary Genius of Plants: A New Understanding of Plant Intelligence and Behavior* (New York: Simon and Schuster, 2018).

30. Camilo Mora, Derek P. Tittensor, Sina Adl, Alastair G. B. Simpson, and Boris Worm, "How Many Species Are There on Earth and in the Ocean?" *PLOS Biology*, August 23, 2011, e1001127; Tod F. Stuessy, "Challenges Facing Systematic Biology," *Taxon* 69, no. 4 (2020): 655–667.

31. Raven et al., "The Origins of Taxonomy."

32. Staffan Müller-Wille, "The Love of Plants," *Nature* 446, no. 7133 (2007): 268.

33. Stefano Mancuso, *The Nation of Plants* (New York: Other Press, 2021): 127.

34. Sune Borkfelt, "What's in a Name? Consequences of Naming Non-Human Animals," *Animals* 1, no. 1 (2011): 116–125.

35. Gledhill, *The Names of Plants*.

36. Phillip V. Tobias, "The Life and Work of Linnaeus," *South African Journal of Science* 74, no. 12 (1978): 457.

37. Endersby, *Imperial Nature*.

38. Müller-Wille, "The Love of Plants."

39. Marc Ereshefsky, *The Poverty of the Linnaean Hierarchy: A Philosophical Study of Biological Taxonomy* (New York: Cambridge University Press, 2000).

40. Endersby, *Imperial Nature*, 178.

41. Quoted in Müller-Wille, "The Love of Plants."

42. Gledhill, *The Names of Plants*.

43. Greta LaFleur, *The Natural History of Sexuality in Early America* (Baltimore: Johns Hopkins University Press, 2018).

44. Daniel R. Headrick, *When Information Came of Age: Technologies of Knowledge in the Age of Reason and Revolution, 1700–1850* (New York: Oxford University Press, 2000).

45. Amy King, *Bloom: The Botanical Vernacular in the English Novel* (New York: Oxford University Press, 2003).

46. Ann B. Shteir, "Sensitive, Bashful, and Chaste? Articulating the Mimosa in Science," in *Science in the Marketplace: Nineteenth-Century Sites and Experiences*, ed. Aileen Fyfe and Bernard Lightman (Chicago: University of Chicago Press, 2007): 169–95; Ann B. Shteir, "*Flora primavera* or *Flora meretrix*? Iconography, Gender, and Science," *Studies in Eighteenth-Century Culture* 36, no. 1 (2007): 147–168.

47. King, *Bloom*; Endersby, *Imperial Nature*.

48. James Prosek, "A Botanist in Swedish Lapland," *New York Times*, May 16, 2017.

49. Lisbet Koerner, *Linnaeus: Nature and Nation* (Cambridge, MA: Harvard University Press, 2001).

50. King, *Bloom*.

51. King, *Bloom*.

52. Bowker, "The Game of the Name."

53. Michel Laurin, "The Subjective Nature of Linnaean Categories and Its Impact in

Evolutionary Biology and Biodiversity Studies," *Contributions to Zoology* 79, no. 4 (October 2010): 131–46; Borkfelt, "What's in a Name?"

54. King, *Bloom.*

55. Borkfelt, "What's in a Name?"

56. Stephen Greenblatt, *Marvelous Possessions: The Wonder of the New World* (Oxford, UK: Clarendon Press, 1991): 80.

57. Borkfelt, "What's in a Name?"

58. Prosek, "A Botanist in Swedish Lapland."

59. Borkfelt, "What's in a Name?"

60. Prosek, "A Botanist in Swedish Lapland."

61. Staffan Müller-Wille, "Nature as a Marketplace: The Political Economy of Linnaean Botany," *History of Political Economy* 35, no. 5 (2003): 154–172.

62. Müller-Wille, "Nature as a Marketplace."

63. Koerner, *Linnaeus.*

64. Koerner, *Linnaeus*, 59.

65. Quoted in Prosek, "A Botanist in Swedish Lapland."

66. Koerner, *Linnaeus*, 7.

67. Geoff Bil, "Imperial Vernacular: Phytonymy, Philology and Disciplinarity in the Indo-Pacific, 1800–1900," *British Journal for the History of Science* 51, no. 4 (December 2018): 635–658; quote from page 657.

68. Quoted in Bowker, "The Game of the Name."

69. Harald Fischer-Tiné and Michael Mann, eds., *Colonialism as Civilizing Mission: Cultural Ideology in British India* (London: Anthem Press, 2004).

70. Stuessy, "Challenges Facing Systematic Biology."

71. E. O. Wilson quoted in Michael Dietrich, "Paradox and Persuasion: Negotiating the Place of Molecular Evolution within Evolutionary Biology," *Journal of the History of Biology* 31 (1998): 85–111.

72. For a wonderful summary of this period see Dietrich, "Paradox and Persuasion."

73. Jack King and Thomas Jukes, "Non-Darwinian Evolution," *Science* 164 (1969): 788–798.

74. Hallam Stevens, *Life Out of Sequence: A Data-Driven History of Bioinformatics* (Chicago: University of Chicago Press, 2019).

75. Michael G. Simpson, *Plant Systematics* (Cambridge, MA: Academic Press, 2019).

76. Bill Bryson, *A Short History of Nearly Everything* (New York: Crown, 2003).

77. Ben Crair, "The Bizarre Bird That's Breaking the Tree of Life," *New Yorker*, July 15, 2022.

78. Quoted in Crair, "The Bizarre Bird."

79. Endersby, "Descriptive and Prescriptive Taxonomies."

80. I draw the ideas of this paragraph from a wonderful blog by Stacey Smith, who succinctly describes cladistic analyses and warns against some common mistakes. Thanks to Madelaine Bartlett for pointing me to it: https://for-the-love-of-trees.blogspot.com/2016/09/the-ancestors-are-not-among-us.html.

81. Geoffrey C. Bowker and Susan Leigh Starr, *Sorting Things Out: Classification and Its Consequences* (Cambridge, MA: MIT Press, 1999).

82. Bowker, "The Game of the Name."

83. Gledhill, *The Names of Plants.*

84. Gledhill, *The Names of Plants.*

85. Gledhill, *The Names of Plants.*

86. Bowker, "The Game of the Name."

87. Gerry Moore, Gideon F. Smith, Estrela Figueiredo, Sebsebe Demissew, Gwilym Lewis, Brian Schrire, Lourdes Rico, and Abraham E. van Wyk, "Acacia, the 2011 Nomenclature Section in Melbourne, and Beyond," *Taxon* 59, no. 4 (August 2010): 1188–1195.

88. K. Montgomery, "Editorial: Brand Acacia Goes to Australia," *South Africa Gardening*, July 4, 2006.

89. Quoted in Jane Carruthers and Libby Robin, "Taxonomic Imperialism in the Battles for Acacia: Identity and Science in South Africa and Australia," *Transactions of the Royal Society of South Africa* 65, no. 1 (2010): 48–64.

90. Quoted in Christian A. Kull and Haripriya Rangan, "Science, Sentiment and Territorial Chauvinism in The Acacia Name Change Debate," *Terra Australis* 34 (2012): 198.

91. Bowker, "The Game of the Name."

92. Kull and Rangan, "Science, Sentiment," 198.

93. Stefan Helmreich, "How Scientists Think; about 'Natives,' for Example. A Problem of Taxonomy among Biologists of Alien Species in Hawaii," *Journal of the Royal Anthropological Institute* 11, no. 1 (2005): 107–128.

94. Quoted in Geoffrey C. Bowker, "Biodiversity Datadiversity," *Social Studies of Science* 30, no. 5 (October 2000): 643–683; quote from 645.

95. Bowker, "Biodiversity Datadiversity," 645.

96. Stuessy, "Challenges Facing Systematic Biology."

97. On the Barcode of Life, see John W. Kress, "Plant DNA Barcodes: Applications Today and in the Future," *Journal of Systematics and Evolution* 55, no. 4 (July 2017): 291–307; Mark Y. Stoeckle and Paul D. N. Hebert, "Barcode of Life," *Scientific American* 299, no. 4 (October 2008): 82–89; and Laura Jones et al., "Barcode UK: A Complete DNA Barcoding Resource for the Flowering Plants and Conifers of the United Kingdom," *Molecular Ecology Resources* 21, no. 6 (2021): 2050–2062. On the Encyclopedia of Life, see Cynthia S. Parr et al., "The Encyclopedia of Life v2: Providing Global Access to Knowledge about Life on Earth," *Biodiversity Data Journal* 2 (2014).

98. Ricki Lewis, "Inventory of Life," *The Scientist* 15, no. 15 (2001): 1.

99. Bowker, "Biodiversity Datadiversity."

100. Bowker, "Biodiversity Datadiversity."

101. Elizabeth Kolbert, *The Sixth Extinction: An Unnatural History* (London: Bloomsbury/A&C Black, 2014).

102. Rouhan and Gaudeul, "Plant Taxonomy."

103. Posey 1997; Helen Watson-Verran and D. Turnbull, "Science and Other Indigenous Knowledge Systems," in *Handbook of Science and Technology Studies*, ed. Sheila Jasanoff, Gerald E. Markle, James C. Petersen, and Trevor Pinch (London: Sage Publications 1994); Bowker, "Biodiversity Datadiversity,"; and Hayden, *When Nature Goes Public.*

104. Bowker, "Biodiversity Datadiversity," 673.

105. Li, "Decolonizing Botanical Genomics."

106. Cyrille de Klemm, *Wild Plant Conservation and the Law* (Gland, Switzerland: International Union for Conservation of Nature and Natural Resources, 1990).

107. Abena Dove Osseo-Asare, *Bitter Roots: The Search for Healing Plants in Africa* (Chicago: University of Chicago Press, 2014).

108. Laura A. Foster, *Reinventing Hoodia: Peoples, Plants, and Patents in South Africa* (Seattle: University of Washington Press, 2017).

109. Figueiredo and Smith, "The Colonial Legacy."

110. Judith E. Winston, "Twenty-First Century Biological Nomenclature: The Enduring Power of Names," *Integrative and Comparative Biology* 58, no. 6 (2018): 1122–1131.

111. Figueiredo and Smith, "The Colonial Legacy."

112. Figueiredo and Smith, "The Colonial Legacy."

113. Figueiredo and Smith, "The Colonial Legacy."

114. Gideon F. Smith and Estrela Figueiredo, "'*Rhodes-*' Must Fall: Some of the Consequences of Colonialism for Botany and Plant Nomenclature," *Taxon* 71, no. 1 (2022): 1–5.

115. Betsy Hartmann, "Conserving Racism: The Greening of Hate at Home and Abroad," *ZNet*, December 10, 2003, http://www.coloursofresistance.org/361/conserving-racism-the-greening-of-hate-at-home-and-abroad/.

116. Dino Grandoni, "National Audubon Society, Pressured to Drop Enslaver's Name, Keeps It," *Washington Post*, March 15, 2023, 2–23.

117. American Ornithological Society, "English Bird Names Project," https://americanornithology.org/about/english-bird-names-project/, accessed December 7, 2023.

118. Gideon F. Smith, Estrela Figueiredo, and Gerry Moore, "Who Amends the International Code of Botanical Nomenclature?" *Taxon* 59, no. 3 (2010): 930–934.

119. Raven et al., "The Origins of Taxonomy."

120. Minakshi Menon, "Making Useful Knowledge: British Naturalists in Colonial India, 1784–1820" (PhD dissertation, UC San Diego).

121. B. T. Styles, "Sir William Jones' Names of Indian Plants," *Taxon* 25, no. 5–6 (1976): 671–674; quote from p. 671.

122. Thanks to Minakshi Menon for this insight. Kathleen Gutierrez makes a similar point in the context of the Philippines in *Sovereign Vernaculars: (Un)Making Botany in the Colonial Philippines* (Durham, NC: Duke University Press, forthcoming).

123. Styles, "Sir William Jones' Names of Indian Plants."

124. K. Eric Wommack and Rita R. Colwell, "Virioplankton: Viruses in Aquatic Ecosystems," *Microbiology and Molecular Biology Reviews* 64, no. 1 (2000): 69–114.

125. Judith Winston, "Systematics and Marine Conservation," in *Systematics, Ecology, and the Biodiversity Crisis*, ed. Niles Eldredge (New York: Columbia University Press, 1992): 157.

126. De Klemm, *Wild Plant Conservation and the Law*, 30.

127. Bowker, "Biodiversity, Datadiversity."

128. Wynter, "Unsettling the Coloniality."

129. McKittrick, *Sylvia Wynter*.

130. Anne McClintock, *Imperial Leather: Race, Gender and Sexuality in the Colonial Contest* (New York: Routledge, 1995), 45.

131. Richard Harry Drayton, "Imperial Science and a Scientific Empire: Kew Gardens and the Uses of Nature, 1772–1903" (PhD dissertation, Yale University, 1993); McClintock, *Imperial Leather*; Philip, *Civilizing Natures*; Michael Goldman, *Imperial Nature: The World Bank and Struggles for Social Justice in the Age of Globalization* (New Haven, CT: Yale University Press, 2005); S. Ravi Rajan, *Modernizing Nature: Forestry and Imperial Eco-Development 1800–1950* (New York: Oxford University Press, 2006); William J. T. Mitchell, "Imperial Landscape," in *The Cultural Geography Reader*, ed. Timothy Oakes and Patricia L. Price (New York: Routledge, 2008), 177–182; John M. MacKenzie, "The Empire of Nature: Hunting, Conservation and British Imperialism," in *The Empire of Nature* (Manchester, UK: Manchester University Press, 2017); Endersby, *Imperial Nature*.

132. Betty V. Smocovitis, "Humanizing Evolution: Anthropology, the Evolutionary Synthesis, and the Prehistory of Biological Anthropology, 1927–1962," *Current Anthropology* 53, no. S5 (2012): S108–S125, https://www.journals.uchicago.edu/doi/full/10.1086/662617.

133. Subramaniam, *Ghost Stories*.

134. Jenny Reardon, "Race to the Finish," in *Race to the Finish: Identity and Governance in an Age of Genomics* (Princeton, NJ: Princeton University Press, 2009).

135. Lisa Gannett, "Racism and Human Genome Diversity Research: The Ethical Limits of Population Thinking," *Philosophy of Science* 68, no. 3 (2001): S479–S492.

136. M. Ty, "Abolish Species: Notes toward an 'Unfenced Is,' Part I," *The Yearbook of Comparative Literature* 64 (2022): 195–226.

137. Bowker, "Biodiversity Datadiversity."

138. Quoted in Juli Berwald, "The Web of Life," *Aeon*, April 5, 2022.

139. Thanks to Mike Dietrich for this insight.

140. Chris Stringer, "Why We Are Not All Multiregionalists Now," *Trends in Ecology & Evolution* 29, no. 5 (2014): 248–251.

141. Gregory W. Stull, Kasey K. Pham, Pamela S. Soltis, and Douglas E. Soltis, "Deep Reticulation: The Long Legacy of Hybridization in Vascular Plant Evolution," *Plant Journal* 114, no. 4 (May 2023).

142. Emily Singer, "A New Approach to Building the Tree of Life," *Quanta Magazine*, June 4, 2013.

143. Singer, "A New Approach."

CHAPTER FOUR | *Perhaps the World Ends Here*

1. Akhil Gupta, "A Different History of the Present: The Movement of Crops, Cuisines, and Globalization," in *Curried Cultures: Globalization, Food, and South Asia*, ed. Krishnendu Ray and Tulasi Srinivas (Berkeley: University of California Press, 2012).

2. Martha Henriques, "How Spices Changed the Ancient World," BBC, https://www.bbc.com/future/bespoke/made-on-earth/the-flavours-that-shaped-the-world/.

3. Cook, *Matters of Exchange*; Kumar, "Botanical Explorations."

4. Ghosh. *The Nutmeg's Curse*.

5. Giles Milton, *Nathaniel's Nutmeg; or, The True and Incredible Adventures of the Spice Trader Who Changed the Course of History* (New York: Macmillan, 1999).

6. Schiebinger and Swan, *Colonial Botany*.

7. Schiebinger and Swan, *Colonial Botany*.

8. The basic descriptions of the *Hortus* and the TKDL are drawn from Banu Subramaniam and Sushmita Chatterjee, "Translations in Green: Colonialism, Postcolonialism, and the Vegetal Turn," *Configurations* (under review).

9. H. Y. Mohan Ram, "On the English Edition of Van Rheede's *Hortus Malabaricus* by K. S. Manilal (2003)," *Current Science* 89, no. 10 (November 25, 2005), 1672–1680.

10. K. S. Manilal, "Medicinal Plants Described in *Hortus Malabaricus*, the First Indian Regional Flora Published in 1678 and Its Relevance to the People of India Today," *Proceedings of the Internation Seminar on "Multidisciplinary Approaches in Angiospenn Systematics"* (West Bengal: University of Kalyani, 2012).

11. K. S. Manilal, *Van Rheede's* Hortus Malabaricus: *Annotated English Edition*, 12 vols. (Thiruvananthapuram: University of Kerala, 2003).

12. K. S. Manilal, "*Hortus Malabaricus* and the Ethnoiatrical Knowledge of Ancient Malabar," *Ancient Science of Life* 4, no. 2 (October 1984): 96–99.

13. Marian Fournier, "Enterprise in Botany: Van Reede and His *Hortus Malabaricus*—Part I," *Archives of Natural History* 14, no. 2 (June 1987): 123–158.

14. K. S. Manilal and M. Remesh, "An Analysis of the Data on the Medicinal Plants Recorded in *Hortus Malabaricus*," *Samagra: Journal of Centre for Research in Indigenous Knowledge Science and Culture* (CRIKSC) 5–6 (2009–2010): 24–72.

15. Fournier, "Enterprise in Botany Part I."

16. Mohan Ram, "On the English Edition."

17. Mohan Ram, "On the English Edition."

18. Mohan Ram, "On the English Edition."

19. Fournier, "Enterprise in Botany Part I."

20. Marian Fournier, "Enterprise in Botany Part II," *Archives of Natural History* 14, no. 3 (June 1987): 297–338; quote is from 306.

21. Fournier, "Enterprise in Botany Part II," 299.

22. K. S. Manilal, "Medicinal Plants," 559.

23. Kapil Raj, "Beyond Postcolonialism . . . and Postpositivism: Circulation and the Global History of Science," *Isis* 104 (2) 2013: 337–47.

24. Grove, "Indigenous Knowledge."

25. Some recent (unpublished) work challenges this story.

26. Fournier, "Enterprise in Botany Part I."

27. Burton Cleetus, "Subaltern Medicine and Social Mobility: The Experience of the Ezhava in Kerala," *Indian Anthropologist* 37, no. 1 (January–June 2007): 147–172.

28. Cleetus, "Subaltern Medicine and Social Mobility," 150.

29. Anna Winterbottom, "Medicine and Botany in the Making of Madras, 1680–1720," in *The East India Company and the Natural World*, ed. Vinita Damodaran, Anna Winterbottom and Alan Lester (London: Palgrave Macmillan, 2014): 35–58.

30. Cleetus, "Subaltern Medicine and Social Mobility."

31. Joseph Satish,"To Seek God in All Things: The Jesuit Encounter with Botany in India," *Journal of the History of Ideas*, August 27, 2018, https://jhiblog.org/2018/08/27/to-seek-god-in-all-things-the-jesuit-encounter-with-botany-in-india/.

32. Subramaniam, *Holy Science.*

33. Leena Abraham, "Gender, Medicine and Globalisation: The Case of Women Ayurveda Physicians of Kerala, India," *Society and Culture in South Asia* 6, no. 1 (2020): 144–164; quote is from 147–148.

34. In some communities women were sequestered in a separate room, relieved of all household duties, and excluded from places of worship. Interestingly, my grandmother told me that these were the times when she was free to read all day—pleasurable moments!

35. Leena Abraham, "From Vaidyam to Kerala Ayurveda." *The Newsletter*, no. 65 (Autumn 2013): 32–33

36. Projit Bihari Mukharji, *Doctoring Traditions: Ayurveda, Small Technologies, and Braided Sciences* (Chicago: University of Chicago Press, 2016).

37. Prakash, *Another Reason.*

38. David Arnold, *Science, Technology, and Medicine in Colonial India* (New York: Cambridge University Press, 2020).

39. Banu Subramaniam, "Viral Fundamentals: Riding the Corona Waves in India," *Religion Compass* 15, no. 2 (2021): e12386.

40. Manilal and Remesh, "An Analysis of the Data."

41. Mohan Ram, "On the English Edition."

42. Manilal, "*Hortus Malabaricus* and the Ethnoiatrical Knowledge."

43. Kapil Raj, "Thinking without the Scientific Revolution: Global Interactions and the Construction of Knowledge," *Journal of Early Modern History* 21 (2017): 445–458.

44. Ezhavas, while OBCs (Other Backward Classes), are a powerful group. Recent unpublished work suggests that there were Dalit actors in this story as well, now doubly erased in the historical record. History, it would seem, is a constant negotiation with power.

45. Sita Reddy, "Making Heritage Legible: Who Owns Traditional Medical Knowledge?," *International Journal of Cultural Property* 13, no 2 (2006): 161–188; Sita Reddy, "Terroir-ism: Putting Heritage in Its Place," *Sharing Cultures 2013*: 481, https://iris.unibas.it/bitstream/11563/64111/1/SC2013_ebook_ATTI_01.pdf#page=500.

46. Mohan Ram, "On the English Edition."

47. Many thanks to Lukas Rieppel for this insight.

48. Geoff Bil and Jaipreet Virdi, "Special Issue Introduction: Colonial Histories of Plant-Based Pharmaceuticals," *History of Pharmacy and Pharmaceuticals* 63, no. 2 (July 2022)) 134–148; DOI: https://doi.org/10.3368/hopp.63.2.134.

49. Pritha Ghosh and Satadru Palbag, "TKDL: An Answer to Biopiracy in India," *International Ayurvedic Medical Journal* 5, no. 11 (November 2017): 4180–4187; quote is from 4180.

50. Jean-Paul Gaudillière, "An Indian Path to Biocapital? The Traditional Knowledge Digital Library, Drug Patents, and the Reformulation Regime of Contemporary Ayurveda," *East Asian Science, Technology and Society: An International Journal* 8, no. 4 (2014): 391–415.

51. Jean-Paul Gaudillière, "An Indian Path to Biocapital?"

52. N. G. Dhawan, M. Mavai, P. Bishnoi, and R. K. Maheshwari, "DTK of Medicines from Bio-piracy: Its Conscientiousness by TKDL of India," *PharmaTutor* 2016, 4 (4); 13–17.

53. Ibid

54. Pradip Thomas, "Copyright and Copyleft in India: Between Global Agendas and Local Interests," in *The Sage Handbook of Intellectual Property*, ed. Matthew David and Debora Halbert (London: Sage, 2015): 355–369.

55. Shiva, *Biopiracy*.

56. Saikat Sen and Raja Chakraborty, "Revival, Modernization and Integration of Indian Traditional Herbal Medicine in Clinical Practice: Importance, Challenges and Future," *Journal of Traditional and Complementary Medicine* 7, no. 2 (2017): 234–244.

57. This is also true of products in the wellness, cosmetic, and beauty industry. Thanks to Karen Lederer for this insight.

58. TKDL researcher quoted in Gaudillière, "An Indian Path to Biocapital?"

59. Thomas, "Copyright and Copyleft in India."

60. Quoted in Gaudillière, "An Indian Path to Biocapital?," 398.

61. Quoted in Gaudillière, "An Indian Path to Biocapital?," 398.

62. Thomas, "Copyright and Copyleft in India."

63. Thomas, "Copyright and Copyleft in India."

64. Leena Abraham, "Gender, Medicine and Globalisation: The Case of Women Ayurveda Physicians of Kerala, India," *Society and Culture in South Asia* 6, no. 1 (2020): 144–164.

65. Gyan Prakash, *Another Reason: Science and the Imagination of Modern India* (Princeton, NJ: Princeton University Press, 1999); Kumar, "Botanical Explorations."

66. Gaudillière, "An Indian Path to Biocapital?"; Harish Naraindas, "Of Spineless Babies and Folic Acid: Evidence and Efficacy in Biomedicine and ayurvedic Medicine," *Social Science & Medicine* 62, no. 11 (June 2006): 2658–2668.

67. Thomas, "Copyright and Copyleft in India."

68. Mukharji, *Doctoring Traditions*.

69. Rachel Berger, *Ayurveda Made Modern: Political Histories of Indigenous Medicine in North India, 1900–1955* (n.p.: Palgrave Macmillan, 2013); Mukharji, *Doctoring Traditions*.

70. Gaudillière, "An Indian Path to Biocapital?"

71. Subramaniam, *Holy Science*

72. Anil Agrawal, "Dismantling the Divide between Indigenous and Scientific Knowledge," *Development and Change* 26, no. 3 (1995): 413–439; quote is from 428.

73. Mukharji, "Vishalyakarani as Eupatorium Ayapana."

74. Mukharji, "Vishalyakarani as Eupatorium Ayapana," 80.

75. Gaudillière, "An Indian Path to Biocapital?," 394

76. Anna Winterbottom, "Becoming 'Traditional': A Transnational History of Neem and Biopiracy Discourse," *Osiris* 36, no. 1 (2021): 262–283.

77. Winterbottom, "Medicine and Botany."

78. Ashis Nandy, *Science, Hegemony, and Violence: A Requiem for Modernity* (Delhi: Oxford University Press, 1988); Claude Alvares, *Science, Development, and Violence:*

The Revolt Against Modernity (Oxford: Oxford University Press, 1992); Vandana Shiva, "Reductionist Science as Epistemological Violence," in *Science, Hegemony, and Violence: A Requiem for Modernity*, ed. Ashis Nandy (Delhi: Oxford University Press, 1988); Vandana Shiva, *Staying Alive: Women, Ecology, and Development* (London: Zed Books, 1989).

79. Dhruv Raina and Irfan Habib, *Domesticating Modern Science: A Social History of Science and Culture in Colonial India* (New Delhi: Tulika, 2006); Dhruv Raina. "Reconfiguring the Center: The Structure of Scientific Exchanges between Colonial India and Europe," *Minerva* 32, no. 2 (1996): 161–76; Druv Raina, "From West to Non-West? Basilla's Three-Stage Model Revisited," *Science as Culture* 8 (1999): 497–516; Dhruv Raina, *Images and Contexts: The Historiography of Science and Modernity in India* (Delhi: Oxford University Pres, 2003); Kapil Raj, "Colonial Encounters and the Forging of New Knowledge and National Identities: Great Britain and India, 1760–1850," *Osiris* 15 (2001): 119–34; Kapil Raj, "Networks of Knowledge, or Spaces of Circulation? The Birth of British Cartography in Colonial South Asia in the Late Eighteenth Century," *Global Intellectual History* 2, no. 1 (2017): 49–66.

80. Nandy, *Science, Hegemony, and Violence.*

81. Gail Omvedt, "Peasants, Dalits and Women: Democracy and India's New Social Movements," *Journal of Contemporary Asia* 24, no. 1 (1994): 35–48.

82. Yogesh Kabirdoss and Samhati Mahapatra, "Tamil Nadu on August 15, 1947: Euphoria and Boycott," Times of India, August 16, 2016, https://timesofindia.indiatimes.com/blogs/tracking-indian-communities/tamil-nadu-on-august-15-1947-euphoria-and-boycott/.

83. PTI, "More Than 19,500 Mother Tongues spoken in India: Census," *Indian Express*, July 1, 2018.

84. George Abraham, "One Nation under Hindutva Speaking Hindi," *India Abroad*, March 19, 2019.

85. Ammon Shea, *Bad English: A History of Linguistic Aggravation* (New York: Penguin, 2014).

86. Yogita Goyal, "Translation and Its Discontents," *Novel: A Forum on Fiction* 52, no. 2 (2019), DOI: 10.1215/00295132-7547074.

87. Arundhati Roy, "What Is the Morally Appropriate Language in Which to Think and Write?," W. G. Sebald Lecture on Literary Translation, June 5, 2018, https://lithub.com/what-is-the-morally-appropriate-language-in-which-to-think-and-write/.

88. Minakshi Menon, "What Is Indian Spikenard?," *South Asian History & Culture* 13, no. 1 (2022): 87–111.

89. Helmreich, "How Scientists Think"; Gutierrez, *Sovereign Vernaculars*.

90. Kumar, "Botanical Explorations."

91. Charu Singh, "Science in the Vernacular? Translation, Terminology and Lexicography in the Hindi Scientific Glossary (1906)," *South Asian History and Culture* 13, no. 1 (2022): 63–86.

92. Some historians argue that phrases like "circulation of knowledge during colonial times" depoliticize the project of colonialism as a form of free trade. It is critical that we not do this. Thanks to Lukas Rieppel for this warning.

93. Li, "Decolonizing Botanical Genomics."

94. Amit Prasad, *Imperial Technoscience: Transnational Histories of MRI in the United States, Britain, and India* (Cambridge, MA: MIT Press, 2014); Philip, *Civilizing Natures.*

95. Audre Lorde, "Uses of the Erotic: The Erotic as Power," in *Sister Outsider*.

96. Cameron Awkward-Rich, *The Terrible We: Thinking with Trans Maladjustment* (Durham, NC: Duke University Press, 2022).

INTERLUDE | *An Ordinary Botany*

1. Katherine McKittrick, *Dear Science and Other Stories* (Durham, NC: Duke University Press, 2020).

CHAPTER FIVE | *The Orchid's Wet Dream*

1. Donna Haraway, *Simians, Cyborgs, and Women: The Reinvention of Nature* (New York: Routledge, 1991); Emily Martin, *The Woman in the Body: A Cultural Analysis of Reproduction* (Boston: Beacon Press, 2001).

2. For an accessible summary see M. Elizabeth Barnes, "'The Spandrels of San Marco and the Panglossian Paradigm: A Critique of the Adaptationist Programme' (1979), by Stephen J. Gould and Richard C. Lewontin," November 14, 2014, https://embryo.asu.edu/pages/spandrels-san-marco-and-panglossian-paradigm-critique-adaptationist-programme-1979-stephen-j.

3. Stephen Jay Gould and Richard C. Lewontin, "The Spandrels of San Marco and the Panglossian Paradigm: A Critique of the Adaptationist Programme," *Proceedings of the Royal Society of London*, Series B: *Biological Sciences* 205, no. 1161 (1979): 581–598; quote is from 582.

4. Stephen J. Gould and Elizabeth S. Vrba, "Exaptation—A Missing Term in the Science of Form," *Paleobiology* 8, no. 1 (1982): 4–15.

5. LaFleur, *The Natural History of Sexuality*.

6. Since the publication of her important book, a growing literature has built on it. I focus on the main story. For further details, see Sally Markowitz, "Pelvic Politics: Sexual Dimorphism and Racial Difference," *Signs* 26, no. 2: (2001): 389–414; Patricia Fara, *Sex, Botany and Empire: The Story of Carl Linnaeus and Joseph Banks* (Icon Books, 2004); Lincoln Taiz and Lee Taiz, *Flora Unveiled: The Discovery and Denial of Sex in Plants* (Oxford University Press, 2017); Sam George, "Botany, Sexuality and Women's Writing, 1760–1830: From Modest Shoot to Forward Plant," in *Botany, Sexuality and Women's Writing, 1760–1830* (Manchester, UK: Manchester University Press, 2017); Melissa Bailes, "Transformations of Gender and Race in Maria Riddell's Transatlantic Biopolitics," *Eighteenth-Century Fiction* 32, no. 1 (2019): 123–144.

7. Schiebinger, *Nature's Body*, 1.

8. Londa Schiebinger, *The Mind Has No Sex? Women in the Origins of Modern Science* (Cambridge, MA: Harvard University Press, 1991), 20–21.

9. Schiebinger, *Nature's Body*.

10. Quoted in Schiebinger, *The Mind Has No Sex?*, 20–21.

11. Londa Schiebinger, "The Loves of the Plants," *Scientific American* 274, no. 2 (1996): 110–115.

12. Maja Bondestam, "When the Plant Kingdom Became Queer: On Hermaphrodites and the Linnaean Language of Nonnormative Sex," in *Illdisciplined Gender: Engaging Questions of Nature/Culture and Transgressive Encounters* , ed. Jacob Bull and Margaretha Fahlgren (Cham, Switzerland: Springer, 2016), 115–135.

13. Schiebinger, *Nature's Body*, 17.

14. Bondestam, "When the Plant Kingdom Became Queer."

15. Schiebinger, *Nature's Body*.

16. Janet Browne, "Botany for Gentlemen: Erasmus Darwin and The Loves of the Plants," *Isis* 80, no. 4 (1989): 592–621.

17. Schiebinger, *Nature's Body*, 22. While human sexuality was projected onto plant and animal worlds, western science's continued erasure of female sexuality is interesting. See especially Rachel E. Gross, *Vagina Obscura: An Anatomical Voyage* (New York: W. W. Norton & Company, 2022).

18. Browne, "Botany for Gentlemen"; Shteir, "Gender and 'Modern' Botany"; Shteir, "Sensitive, Bashful, and Chaste"; George, "Linnaeus in Letters."

19. George, "Linnaeus in Letters."

20. Paula Findlen, review of *Nature's Body*, "The J. H. B. Bookshelf," *Journal of the History of Biology* 28 (1995): 369–379.

21. Schiebinger, *Nature's Body*, 23.

22. Quoted in Fara, *Sex, Botany and Empire*, 19.

23. Ernst Mayr, *The Growth of Biological Thought: Diversity, Evolution, and Inheritance* (Cambridge, MA: Belknap Press, 1982).

24. Michelle Murphy, *Seizing the Means of Reproduction* (Durham, NC: Duke University Press, 2012).

25. Headrick, *When Information Came of Age*.

26. Headrick, *When Information Came of Age*.

27. Staffan Müller-Wille, "Linnaeus and the Four Corners of the World," in *The Cultural Politics of Blood, 1500–1900*, ed. Kimberly Anne Coles, Ralph Bauer, Zita Nunes and Carla L. Peterson (Basingstoke: Palgrave MacMillan, 2015), 191–209.

28. Schiebinger, *Nature's Body*; Markowitz, "Pelvic Politics."

29. Headrick, *When Information Came of Age*; Monica Gagliano, John C. Ryan, and Patricia Vieira, eds., *The Language of Plants: Science, Philosophy, Literature* (Minneapolis: University of Minnesota Press, 2017).

30. Browne, "Botany for Gentlemen."

31. Vandana Shiva, *Monocultures of the Mind: Perspectives on Biodiversity and Biotechnology* (New York: Zed Books, 1993).

32. Jamaica Kincaid, *My Garden (Book)* (New York: Farrar, Straus and Giroux, 1999), 166.

33. Mary Louise Pratt, "Science, Planetary Consciousness, Interiors," in *Imperial Eyes: Travel Writing and Transculturation* (New York: Routledge, 1992).

34. Shteir, "Gender and 'Modern' Botany," 29.

35. George, "Botany, Sexuality and Women's Writing."

36. Nancy F. Cott, "Passionlessness: An Interpretation of Victorian Sexual Ideology, 1790–1850," *Signs: Journal of Women in Culture and Society* 4, no. 2 (1978): 219–236; Mark T. Hoyer, "Cultivating Desire, Tending Piety: Botanical Discourse in Harriet Beecher Stowe's *The Minister's Wooing,*" *Beyond Nature Writing: Expanding the Boundaries of Ecocriticism* (2001): 111–125.

37. Schiebinger, *Nature's Body.*

38. Quoted in Bondestam, "When the Plant Kingdom Became Queer," 123. Also see Rogel L. Williams, *Botanophilia in Eighteenth-Century France: The Spirit of the Enlightenment* (Boston: Kluwer Academic Publishers, 2001), 179.

39. Sam George, "Forward Plants and Wanton Women: Botany and Sexual Anxiety in the Late Eighteenth Century," in *Botany, Sexuality and Women's Writing, 1760–1830* (Manchester, UK: Manchester University Press, 2017), 105–152.

40. While I am homogenizing the impact of Linnaeus across the western world, there were differences across nations. For a thicker description of this history, see George, "Botany, Sexuality and Women's Writing"; Schiebinger, *The Mind Has No Sex*; Browne, "Botany for Gentlemen"; Barbara Gates, *Kindred Nature: Victorian and Edwardian Women Embrace the Living World* (Chicago: University of Chicago Press 1998); Shteir, *Cultivating Women*; Tim Fulford, "Coleridge, Darwin, Linnaeus: The Sexual Politics of Botany," *The Wordsworth Circle* 28, no. 3 (1997): 124–130; Alan Bewell, "Jacobin Plants: Botany as Social Theory in the 1790s," *The Wordsworth Circle* 20, no. 3 (1989): 132–139.

41. Sam George, "Carl Linnaeus, Erasmus Darwin and Anna Seward: Botanical Poetry and Female Education," *Science & Education* 23, no. 3 (2014); 673–694.

42. "The Legacy of Linnaeus" (editorial), *Nature* 446 (March 15, 2007): 231–232.

43. For other terms, see "Plant Reproductive Morphology," *Wikipedia*, https://en.wikipedia.org/wiki/Plant_reproductive_morphology.

44. Evelyn Fox Keller, "On the Need to Count Past Two in our Thinking about Gender and Science," *New Ideas in Psychology* 5, no. 2 (1987): 275–287.

45. Bondestam, "When the Plant Kingdom Became Queer." Greta LaFleur, *The Natural History of Sexuality in Early America* (Baltimore: Johns Hopkins University Press, 2020) also documents mixed-sex radical political and intellectual groups, like the Friendly Club in New York, that read Linnaeus alongside radical political theorists of the day.

46. Elizabeth Grosz, *Becoming Undone: Darwinian Reflections on Life, Politics, and Art* (Durham, NC: Duke University Press, 2011); Carla Hustak and Natasha Myers, "Involutionary Momentum: Affective Ecologies and the Sciences of Plant/Insect Encounters," *differences* 23, no. 3 (2012): 74–118; Elizabeth Wilson, "Biologically Inspired Feminism: Response to Helen Keane and Marsha Rosengarten, 'On the Biology of Sexed Subjects,'" *Australian Feminist Studies* 17, no. 39 (2002): 283–285; D. O. Schaefer, "Darwin's Orchids: Evolution, Natural Law, and the Diversity of Desire," *GLQ* 27, no. 4 (2021: 525–550).

47. Thanks to Beans Velocci for a productive conversation on this point.

48. Bill Finch, "Why Plants Have Sex, and Why It Matters to Your Garden," AL.com (Alabama), July 22, 2011: https://www.al.com/living-press-register/2011/07/why_plants_have_sex_and_why_it.html.

49. Banu Subramaniam and Madelaine Bartlett, "Re-imagining Reproduction: The Queer Possibilities of Plants," *Integrative and Comparative Biology* 63, no. 4 (October 2023), icad012, https://doi.org/10.1093/icb/icad012.

50. Donald R. Kaplan and Wolfgang Hagemann, "The Relationship of Cell and Organism in Vascular Plants: Are Cells the Building Blocks of Plant Form?," *BioScience* 41 (1991): 693–703.

51. Paco Garcia-Gonzalez, Damian Dowling, and Magdalena Nystrand, "Male, Female—Ah, What's the Difference?," *The Conversation*, March 26, 2013.

52. Michael Breed and Janice Moore, "Evolution of Sex: Why Some Animals Are Called Male and Others Female," in *Animal Behavior*, 2nd ed. Amsterdam:Elsevier, 2016): 360–365.

53. Anne Fausto-Sterling, "Feminism and Behavioral Evolution: A Taxonomy," in *Feminism and Evolutionary Biology*, ed. Patricia Adair Gowaty (Boston, MA: Springer, 1997), 42–60.

54. George C. Williams, *Sex and Evolution* (Princeton, NJ: Princeton University Press, 1975).

55. Fausto-Sterling, "Feminism and Behavioral Evolution."

56. Michael R. Whitehead, Robert Lanfear, Randall J. Mitchell, and Jeffrey D. Karron, "Plant Mating Systems Often Vary Widely among Populations," *Frontiers in Ecology and Evolution* 6 (2018): 38.

57. The evolution of sex remains one of the hotly debated questions in biology.

58. Quentin Cronk, "Some Sexual Consequences of Being a Plant," *Philosophical Transactions of the Royal Society B* 377, no. 1850 (2022): p.20210213.

59. E. Albertini, G. Barcaccia, J. G. Carman, and F. Pupilli, "Did Apomixis Evolve from Sex or Was It the Other Way Around?" *Journal of Experimental Botany* 70, no. 11 (2019): 2951–2964.

60. A. J. Bateman, "Intra-Sexual Selection in Drosophila," *Heredity* 2, no. 3 (1948): 349–368.

61. Marta L. Wayne, "Walking a Tightrope: The Feminist Life of a Drosophila Biologist," NWSA *Journal* 12, no. 3 (2000): 139–50; quote from page 140.

62. Bateman, "Intra-Sexual Selection in Drosophila."

63. Zuleyma Tang-Martínez, "Rethinking Bateman's Principles: Challenging Persistent Myths of Sexually Reluctant Females and Promiscuous Males," *Journal of Sex Research* 53, no. 4–5 (2016): 532–559; Zuleyma Tang-Martinez and T. B. Ryder, "The Problem with Paradigms: Bateman's Worldview as a Case Study," *Integrative and Comparative Biology* 45, no. 5 (2005): 821–830; P. G. Parker and Zuleyma Tang-Martinez, "Bateman Gradients in Field and Laboratory Studies: A Cautionary Tale," *Integrative and Comparative Biology* 45, no. 5 (2005): 895–902; Sarah B. Hrdy and Ruth Bleier, "Empathy, Polyandry, and the Myth of the Coy Female," *Conceptual Issues in Evolutionary Biology* 131 (1986); Sarah B. Hrdy, *The Woman That Never Evolved* (Cambridge, MA: Harvard University Press, 2009); Patricia Adair Gowaty and Stephen P. Hubbell, "Chance, Time Allocation, and the Evolution of Adaptively Flexible Sex Role Behavior," *Integrative and Comparative Biology* 45, no. 5 (2005): 931–944; Brian F. Snyder and Patricia Adair Gowaty, "A Reappraisal of

Bateman's Classic Study of Intrasexual Selection," *Evolution: International Journal of Organic Evolution* 61, no. 11 (2007): 2457–2468; Patricia Adair Gowaty, Yong-Kyu Kim, and Wyatt W. Anderson, "No Evidence of Sexual Selection in a Repetition of Bateman's Classic Study of *Drosophila melanogaster*," *Proceedings of the National Academy of Sciences* 109, no. 29 (2012): 11740–11745; Thierry Hoquet, William C. Bridges, and Patricia Adair Gowaty, "Bateman's Data: Inconsistent with 'Bateman's Principles,'" *Ecology and Evolution* 10, no. 19 (2020): 10325–10342.

64. Myrna Perez Sheldon, "Sexual Selection as Race Making," *BJHS Themes* 6 (2012): 9–23.

65. Jeanne Tonnabel, Patrice David, and John R. Pannell, "Do Metrics of Sexual Selection Conform to Bateman's Principles in a Wind-Pollinated Plant?," *Proceedings of the Royal Society B* 286, no. 1905 (2019): 20190532.

66. M. Shrestha, A. G., Dyer, A. Dorin, Z. X. Ren, and M. Burd, "Rewardlessness in Orchids: How Frequent and How Rewardless?," *Plant Biology* 22, no. 4 (2020): 555–561.

67. Michel Baguette, Joris A. Bertrand, Virginie Marie Stevens, and Bertrand Schatz, "Why Are There So Many Bee-Orchid Species? Adaptive Radiation by Intra-Specific Competition for Mnesic Pollinators," *Biological Reviews* 95, no. 6 (2020), 1630–1663.

68. Michael Pollan, "The Weird Sex Life of Orchids," *The Guardian*, October 8, 2011.

69. Quoted in Pollan, "The Weird Sex Life of Orchids."

70. They invite the bee deep into the flower. In "Orchids: Nature's Sexy Sirens" (https://www.abc.net.au/science/articles/2008/09/25/2373397.htm), Heather Catchpole argues, "Orchids differ from other flowers by holding both their male and female parts in a column at the base of the flower. Their three sepals and petals have evolved to look like a labellum or lip. The lip acts as a visual cue, landing platform, and occasionally a trap for insect pollinators. Orchids employ several different 'lip conversions' to lure insects. These can include ornaments, such as pseudo anthers, hinting there's lots of pollen on offer; sexual lures, where the lip mimics a female insect of its pollinator species; a glossy appearance, as if the petal is dripping with nectar; and UV spectrum cues, like spots and lines, which act like runway lights to guide insects into the flower.

71. There is an important literature on settler colonialism and sexuality. See Kim TallBear, "Making Love and Relations beyond Settler Sex and Family," in *Making Kin Not Population*, ed. Adele Clarke and Donna Haraway (Chicago: Prickly Paradigm Press, 2018); Kim TallBear and Angela Willey, "Critical Relationality: Queer, Indigenous, and Multispecies Belonging beyond Settler Sex and Nature," *Imaginations: Journal of Cross-Cultural Image Studies* 10, no. 1 (2019): 5–15.

72. Myra Hird, "Animal Transex," *Australian Feminist Studies* 21, no. 49 (March 2006): 35–50.

73. Katherine McKittrick, *Dear Science and Other Stories* (Durham, NC: Duke University Press, 2020).

74. Subramaniam, *Holy Science*.

75. Ashis Nandy, *The Intimate Enemy: Loss and Recovery of Self under Colonialism* (Oxford: Oxford University Press, 1989).

76. Pratt, "Science, Planetary Consciousness, Interiors."

77. Roughgarden, *Evolution's Rainbow*; Joan Roughgarden, "The Myth of Sexual Selection," *California Wild: The Magazine of the California Academy of Sciences*, Summer, 2005; Joan Roughgarden, *The Genial Gene: Deconstructing Darwinian Selfishness* (Berkeley: University of California Press, 2009).

78. See papers from the "Sexual Diversity and Variation" issue of *Integrative and Comparative Biology Journal* 63, no. 4 (October 2023).

79. Lynn Margulis and Dorian Sagan, *Origins of Sex: Three Billion Years of Genetic Recombination* (New Haven, CT: Yale University Press, 1986).

80. Fausto-Sterling, "Feminism and Behavioral Evolution."

81. Catriona Sandilands, "Queer Ecology," in *Keywords for Environmental Studies*, ed. Joni Adamson, William A. Gleason, and David N. Pellow (New York: New York University Press, 2016), 169–171; quote is from 171.

82. Thanks to Beans Velocci for pushing me to challenge assumptions of queer sexuality.

83. Subramaniam, *Ghost Stories*.

84. Wynter, "Unsettling the Coloniality"; Gloria Anzaldúa, *Borderlands: The New Mestiza* (San Francisco: Aunt Lute Books, 1987); Willey, *Undoing Monogamy*; McKittrick, *Dear Science*; Herzig, *Plucked*; Donna Haraway, *Staying with the Trouble: Making Kin in the Chthulucene* (Durham, NC: Duke University Press, 2016); Mohanty, *Feminism without Borders: Decolonizing Theory, Practicing Solidarity*; (Durham: Duke University Press 2003); José Esteban Muñoz, *Disidentifications: Queers of Color and the Performance of Politics*, vol. 2 (Minneapolis: University of Minnesota Press, 1999).

85. Wilson, "Biologically Inspired Feminism"; Noreen Giffney and Myra J. Hird, eds., *Queering the Non/Human* (Burlington, VT: Ashgate, 2008); Grosz, *Becoming Undone*; Myra J. Hird, "Animal Transsex," in *The Transgender Studies Reader* 2, ed. Susan Stryker and Aren Z. Aizura (New York: Routledge, 2013), 156–167; Hustak and Myers, "Involutionary Momentum."

86. Siobhan Somerville, "Scientific Racism and the Emergence of the Homosexual Body," *Journal of the History of Sexuality* 5, no. 2 (1994): 243–266; Hortense Spillers, "Mama's Baby, Papa's Maybe: An American Grammar Book," *Diacritics* 17, no. 2 (1987): 64–81; Samantha Pinto, "Black Feminist Literacies: Ungendering, Flesh, and Post-Spillers Epistemologies of Embodied and Emotional Justice," *Journal of Black Sexuality and Relationships* 4, no. 1 (2017): 25–45.

CHAPTER SIX | *In the Dark Shadows of the Tree of Life*

1. David Margolick, *Strange Fruit: The Biography of a Song* (New York: Ecco Press, 2001).

2. Sarah Prager, "Four Flowering Plants That Have Been Decidedly Queered," *JSTOR Daily*, January 29, 2020, https://daily.jstor.org/four-flowering-plants-decidedly-queered/.

3. Gabriella V. Smith, "Uncle Tom's Garden: Color, Culture, and Racialization in Garden Plants," *Sociological Inquiry* 92, no. 2 (2022): 466–489. Much like Harlan Weaver's powerful analysis of dog breeding and race (*Bad Dog* [Seattle: University of Washington Press, 2021]), a rich analysis of the racial politics of plants and flowers is a project waiting to happen. Many thanks to Angie Willey for raising these connections.

4. Christopher Krupa, *A Feast of Flowers: Race, Labor and Postcolonial Capitalism in Ecuador* (Philadelphia: University of Pennsylvania Press, 2022).

5. Thanks to Great LaFleur for pointing me to Deborah A. Miranda, "Extermination of the *Joyas*: Gendercide in Spanish California," *GLQ: A Journal of Lesbian and Gay Studies* 16, no. 1–2 (2010): 253–284, and Scott L. Morgensen, *Spaces between Us: Queer Settler Colonialism and Indigenous Decolonization* (Minneapolis: University of Minnesota Press, 2011).

6. Lugones, "Heterosexualism and the Colonial/Modern Gender System"; Maria Lugones, "The Coloniality of Gender," in *The Palgrave Handbook of Gender and Development*, ed. Wendy Harcourt (London: Palgrave Macmillan, 2016), 13–33; Wynter, "Unsettling the Coloniality"; Breny Mendoza, "Coloniality of Gender and Power," in *The Oxford Handbook of Feminist Theory*, ed. Lisa Disch and Mary Hawkesworth (New York: Oxford University Press, 2015), 100–121; Walsh, "Shifting the Geopolitics"; Freya Schiwy, "Decolonization and the Question of Subjectivity: Gender, Race, and Binary Thinking," *Cultural studies* 21, no. 2–3 (2007): 271–294; Asher, "Latin American Decolonial Thought."

7. Lugones, "The Coloniality of Gender."

8. Anzaldúa, *Borderlands*, 26.

9. Thanks to Great LaFleur for pointing me to these readings: Jennifer M. Spear, *Race, Sex, and Social Order in Early New Orleans* (Baltimore: Johns Hopkins University Press, 2009); Zeb Tortorici, *Sins against Nature: Sex and Archives in Colonial New Spain* (Durham, NC: Duke University Press, 2018); Lamonte Aidoo, *Slavery Unseen: Sex, Power, and Violence in Brazilian History* (Durham, NC: Duke University Press, 2018).

10. Spillers, "Mama's Baby, Papa's Maybe"; Qwo-Li Driskill, *Asegi Stories: Cherokee Queer and Two-Spirit Memory* (Tucson: University of Arizona Press, 2016); C. Riley Snorton. *Black on Both Sides: A Racial History of Trans Identity* (Minneapolis: University of Minnesota Press, 2017); Jules Gill-Peterson, "Gender," in *Keywords for Gender and Sexuality Studies*, ed. The Keywords Feminist Editorial Collective (New York: NYU Press, 2021).

11. Maria Lugones, "Toward a Decolonial Feminism," *Hypatia* 25, no. 4 (2010): 743.

12. There were many debates on which groups belong to the category human. In short, many racist theories exist about human difference.

13. Thanks to Greta LaFleur for pushing me to use more precise language.

14. LaFleur, *The Natural History of Sexuality*.

15. Including but not limited to Anne Stoler, *Race and the Education of Desire: Foucault's History of Sexuality and the Colonial Order of Things* (Durham, NC: Duke University Press, 1995); Robert J. C. Young, "Foucault on Race and Colonialism," *New Formations*, 1995: 57–57; Siobhan B. Somerville, *Queering the Color Line: Race and the Invention of Homosexuality in American Culture* (Durham, NC: Duke University Press, 2000); A. Cruz-Malavé, M. F. Manalansan, and M. Manalansan, eds., *Queer Globalizations: Citizenship and the Afterlife of Colonialism*, vol. 9 (New York: NYU Press, 2002); Antoinette Burton, *Gender, Sexuality and Colonial Modernities* (New York: Routledge, 2005); Anjali Arondekar, *Abundance: Sexuality's History* (Durham, NC: Duke University Press, 2023); Anne McClintock, *Imperial Leather: Race, Gender, and Sexuality in the*

Colonial Contest (New York: Routledge, 2013); A. G. Weheliye, *Habeas Viscus: Racializing Assemblages, Biopolitics, and Black Feminist Theories of the Human* (Durham, NC: Duke University Press, 2014); Nancy R. Hunt, *A Nervous State: Violence, Remedies, and Reverie in Colonial Congo* (Durham, NC: Duke University Press, 2015); Willey, *Undoing Monogamy*.

16. Lugones, "The Coloniality of Gender," 195.

17. Jennifer Morgan, *Laboring Women: Reproduction and Gender in New World Slavery* (Philadelphia: University of Pennsylvania Press, 2004); LaFleur, *The Natural History of Sexuality*.

18. Anne Fausto-Sterling, "Gender, Race, and Nation," in *Deviant Bodies: Critical Perspectives on Difference in Science and Popular Culture*, ed. Jennifer Terry and Jacqueline Urla (Bloomington: Indiana University Press, 1995); Zine Magubane, "Which Bodies Matter? Feminism, Poststructuralism, Race, and the Curious Theoretical Odyssey of the 'Hottentot Venus,'" *Gender & Society* 15, no. 6 (2001): 816–834; Clifton Crais and Pamela Scully, *Sara Baartman and the Hottentot Venus* (Princeton, NJ: Princeton University Press, 2021).

19. Evelynn Hammonds, "Black (W)holes and the Geometry of Black Female Sexuality," *Differences: A Journal of Feminist Cultural Studies* 6, no. 2–3 (1994): 126–146.

20. Marie Draz, "On Gender Neutrality: Derrida and Transfeminism in Conversation," *philoSOPHIA: A Journal of Continental Feminism* 7, no. 1 (2017): 91–98.

21. Mel Y. Chen, *Animacies* (Durham, NC: Duke University Press, 2012), 137.

22. Zakiyyah Iman Jackson, "Becoming Human," in *Becoming Human* (New York: New York University Press, 2020). Also see Eva Hayward and Che Gossett, "Impossibility of That," *Angelaki* 22, no. 2 (2017): 15–24; Kadji Amin, "Trans* Plasticity and the Ontology of Race and Species," *Social Text* 38, no. 2 (2020): 49–71; Rosalba Icaza and Rolando Vázquez, "The Coloniality of Gender as a Radical Critique of Developmentalism," in *The Palgrave Handbook of Gender and Development*, ed. Wendy Harcourt (London: Palgrave Macmillan, 2016), 62–73.

23. Alexis Pauline Gumbs, *Undrowned: Black Feminist Lessons from Marine Mammals* (Chico, CA: AK Press, 2020). Many thanks to Karen Cardozo for introducing me to this work.

24. Mahmood Mamdani, *Neither Settler nor Native* (Cambridge, MA: Harvard University Press, 2020).

25. María Elena Martínez, *Genealogical Fictions: Limpieza de Sangre, Religion, and Gender in Colonial Mexico* (Stanford, CA: Stanford University Press, 2008).

26. David Penny, "Darwin's Theory of Descent with Modification, versus the Biblical Tree of Life," *PLoS Biology* 9, no. 7 (2011), e1001096.

27. While Darwin's figure shaped biology's history with trees as genealogical representations of life on earth, there is a healthy debate on other interpretations of the figure. For example, see Horst Bredekamp, *Darwin's Corals: A New Model of Evolution and the Tradition of Natural History* (De Gruyter, 2019); János Podani, "The Coral of Life," *Evolutionary Biology* 46, no. 2 (2019): 123–144.

28. Isabelle Charmantier, "Linnaeus and Race," *The Linnaean Society of London*: 2020,

https://www.linnean.org/learning/who-was-linnaeus/linnaeus-and-race. As Müller-Wille argues, "Presenting Linnaeus's distinction as a series of trinomials goes back at least to Stephen Jay Gould's *Mismeasure of Man*, p. 66, and probably has its origin in an English translation of the first part of thirteenth, posthumous edition of Systema Naturae that was published in 1792"; Müller-Wille, "Linnaeus and the Four Corners of the World," note 6.

29. Müller-Wille, "Linnaeus and the Four Corners of the World," note 6.

30. Witmer Stone, "Racial Variation in Plants and Animals, with Special Reference to the Violets of Philadelphia and Vicinity," *Proceedings of the Academy of Natural Sciences of Philadelphia* 55 (1903): 656–699.

31. S. O. Y. Keita, R. A. Kittles, C. D. Royal, G. E. Bonney, P. Furbert-Harris, G. M. Dunston, and C. N. Rotimi, "Conceptualizing Human Variation," *Nature Genetics* 36, S17-S20 (2004).

32. Hull, "Species, Subspecies, and Races."

33. While the term *subspecies* may not be used within population genetics and phylogeography, scientists do attempt to understand the distribution of genetic variation in ecological terms. See John C. Avise, *Phylogeography: The History and Formation of Species* (Cambridge, MA: Harvard University Press, 2000).

34. Roderick A. Ferguson, *Aberrations in Black: Toward a Queer of Color Critique* (Minneapolis: University of Minnesota Press, 2003).

35. LaFleur, *The Natural History of Sexuality*.

36. Thanks to Samantha Pinto for this insight.

37. Schiebinger, *Nature's Body*.

38. For a discussion on indigeneity, polyamory, and settler colonialism see TallBear, "Making Love and Relations."

39. Nancy Stepan, "Race and Gender: The Role of Analogy in Science," *Isis* 77, no. 2 (1986): 261–77. Despite the title, what she is discussing in her article is, per our terminology today, sex, not gender.

40. Stepan, "Race and Gender," 40–41.

41. Markowitz, "Pelvic Politics," 391.

42. Markowitz, "Pelvic Politics," 391.

43. Evelynn Hammonds, "The Logic of Difference: A History of Race in Science and Medicine in the United States," presentation at the Women's Studies Program, UCLA, 1999.

44. Thanks to Samantha Pinto for her insights here. For more information, see Deirdre Cooper Owens, *Medical Bondage: Race, Gender, and the Origins of American Gynecology* (Athens: University of Georgia Press, 2017); Rebecca Wanzo, *This Suffering Will Not Be Televised* (New York: State University of New York Press, 2009).

45. Saidiya Hartmann, "The Hold of Slavery," *New York Review of Books*, October 24, 2022.

46. Laura Briggs, *How All Politics Became Reproductive Politics: From Welfare Reform to Foreclosure to Trump*, Vol. 2 (Berkeley: University of California Press, 2018).

47. Morgan, *Laboring Women*; Shatema Threadcraft, *Intimate Justice: The Black Female Body and the Body Politic* (New York: Oxford University Press, 2016); Alys E. Weinbaum,

The Afterlife of Reproductive Slavery: Biocapitalism and Black Feminism's Philosophy of History (Durham, NC: Duke University Press, 2019).

48. Khiara Bridges, *Reproducing Race: An Ethnography of Pregnancy as a Site of Racialization* (Berkeley: University of California Press, 2011); Roberts, *Killing the Black Body*; Washington, *Medical Apartheid*; D. A. Davis, *Reproductive Injustice* (New York: New York University Press, 2019); Sara Matthiesen, *Reproduction Reconceived* (Berkeley: University of California Press, 2021); Jennifer C. Nash, *Birthing Black Mothers* (Durham, NC: Duke University Press, 2021); Natali N. Valdez, *Weighing the Future: Race, Science, and Pregnancy Trials in the Postgenomic Era* (Berkeley: University of California Press, 2021).

49. Krista Lynes, "Pollination and the Horrors of Yield: Scarcity and Survival in the Glasshouse," *Catalyst* 9, no. 1 (2023): 1–24

50. Cardozo and Subramaniam, "Assembling Asian/American Naturecultures"; Anna L. Tsing, "Empowering Nature, or: Some Gleanings in Bee Culture," in *Naturalizing Power: Essays in Feminist Cultural Analysis*, ed. Sylvia Yanagisako and Carol Delaney (New York: Routledge, 1995), 113–143.

51. Weinbaum, *Afterlife*.

52. Dorothy Roberts, *Killing the Black Body: Race, Reproduction, and the Meaning of Liberty* (New York: Vintage, 2014); Morgan, *Laboring Women*; Shatema Threadcraft, *Intimate Justice*; Weinbaum, *Afterlife*.

53. Clarence C. Gravlee, "Systemic Racism, Chronic Health Inequities, and COVID-19: A Syndemic in the Making?," *American Journal of Human Biology* 32, no. 5 (2020); Rohan Khazanchi, Charlesnika T. Evans, and Jasmine R. Marcelin, "Racism, Not Race, Drives Inequity across the COVID-19 Continuum," *JAMA Network Open* 3, no. 9) (2020), pp.e2019933-e2019933; Elora Halim Chowdhury, "The Precarity of Preexisting Conditions," *Feminist Studies* 46, no. 3 (2020): 615–625; Mel Y. Chen, "Feminisms in the Air," *Signs: Journal of Women in Culture and Society* 47, no. 1 (2021): 22–29; Evelynn M. Hammonds, "A Moment or a Movement? The Pandemic, Political Upheaval, and Racial Reckoning," *Signs: Journal of Women in Culture and Society* 47, no. 1 (2021): 11–14.

54. Kimmerer, *Braiding Sweetgrass*; Vandana Shiva and Kunwar Jalees, *Seeds of Suicide: The Ecological and Human Costs of Seed Monopolies and Globalisation of Agriculture* (New Delhi: Research Foundation for Science, Technology, and Ecology, 2000).

55. Michael M. Pollan, *The Botany of Desire: A Plant's-Eye View of the World* (New York: Random House, 2002).

56. Food and Agriculture Organization of the United Nations, "Once Neglected, These Traditional Crops Our New Rising Stars," February 10, 2018, https://www.fao.org/fao-stories/article/en/c/1154584/.

57. Kimmerer, *Braiding Sweetgrass*, 138.

58. Thanks to Madelaine Bartlett for pushing me to recognize the diversity of agricultural practices.

59. Samantha M. Ohlgart, "The Terminator Gene: Intellectual Property Rights vs. the Farmer's Common Law Right to Save Seeds," *Drake Journal of Agricultural Law* 7 (2002): 473.

60. Shiva et al., *Seeds of Suicide.*

61. Eric Niiler, "Terminator Technology Temporarily Terminated," *Nature Biotechnology* 17, 1 (1999): 1054

62. Canadian Biotechnology Action Network, "Terminator Technology," https://cban.ca/gmos/issues/terminator-technology/, accessed December 10, 2023.

63. Donna Haraway, "A Cyborg Manifesto (1985)," in *Cultural Theory: An Anthology*, ed. Imre Szeman and Timothy Kaposy (Hoboken, NJ: Wiley, 2010): 454.

64. Cedric J. Robinson, *Black Marxism: The Making of the Black Radical Tradition*, rev. and updated 3rd ed. (Chapel Hill: University of North Carolina Press, 2020).

65. Lisa Lowe, *The Intimacies of Four Continents* (Durham, NC: Duke University Press, 2015), 150.

66. Xan S. Chacko, "Stringing, Reconnecting, and Breaking the Colonial 'Daisy Chain': From Botanic Garden to Seed Bank," *Catalyst: Feminism, Theory, Technoscience* 8, no. 1 (2022).

67. Subramaniam, *Holy Science.*

68. Aniket Aga, *Genetically Modified Democracy: Transgenic Crops in Contemporary India* (New Haven, CT: Yale University Press, 2021).

69. Natali Valdez, *Weighing the Future: Race, Science, and Pregnancy Trials in The Postgenomic Era*, vol. 9 (Berkeley: University of California Press, 2021).

70. Banu Subramaniam, "The Ethical Impurative," in *MEAT! A Transnational Analysis*, ed. Sushmita Chatterjee and Banu Subramaniam (Durham, NC: Duke University Press, 2021).

71. B. Glaeser, ed., *The Green Revolution Revisited: Critique and Alternatives*, vol. 2 (Taylor & Francis, 2010); Paul Mosley, "Two Africas? Why Africa's 'Growth Miracle' Is Barely Reducing Poverty," Brooks World Poverty Institute Working Paper 191 (January 2, 2014); Benjamin Robert Siegel, *Hungry Nation: Food, Famine, and the Making of Modern India* (New York: Cambridge University Press, 2018).

72. R. B. Singh, "Environmental Consequences of Agricultural Development: A Case Study from the Green Revolution State of Haryana, India," *Agriculture, Ecosystems & Environment* 82, no. 1–3 (2000): 97–103.

73. Ranjit Singh Ghuman, "Act Now to Save Punjab on the Water Front," *The Tribune*, May 20, 2022.

74. Leanne Betasamosake Simpson, "Temporary Spaces of Joy and Freedom: Leanne Betasmosake Simpson and Dionne Brand," *Literary Review of Canada*, June 2018.

75. On degendering of plants, see Kimmerer, *Braiding Sweetgrass.*

76. Chen, *Animacies*; Kimmerer, *Braiding Sweetgrass.*

77. Rob Nixon, "The Less Selfish Gene: Forest Altruism, Neoliberalism, and the Tree of Life," *Environmental Humanities* 13, no. 2 (2021): 348–371.

78. Lauren Berlant, "Slow Death (Sovereignty, Obesity, Lateral Agency)," *Critical Inquiry* 33, no. 4 (2007): 754–780; quote is from 760.

79. Roberts, *Killing the Black Body.*

80. Yingchen Kwok, "Method and Pedagogy in Trans Studies, Trans History, and the History of Science: An Interview with Beans Velocci," *Penn Arts & Sciences*, April 4, 2022, https://gsws.sas.upenn.edu/news/2022/04/04/interview-beans-velocci-yingchen-kwok

81. Sharon Kinsman, "Life, Sex, and Cells," in *Feminist Science Studies: A New Generation*, ed. Maralee Mayberry, Banu Subramaniam, and Lisa Weasel (New York: Routledge, 2001), 193–203; quote is from 197.

82. Elizabeth Wilson, "Biologically Inspired Feminism: Response to Helen Keane and Marsha Rosengarten, 'On the Biology of Sexed Subjects,'" *Australian Feminist Studies* 17, no. 39 (2002): 283–285.

INTERLUDE | *The Queer Vegennials*

1. Deep gratitude to Angie Willey for teaching me this important mantra.

INTERLUDE | *International Council for Queer Planetarity*

1. Studies have repeatedly established that indigenous communities are better stewards of the land than settler populations. The land is not an inert landscape but a living ecoscape of multitudinous and diverse cosmologies and ecologies.

2. The Council acknowledges colonization of land and ongoing settler colonialism. Many citizens of formerly colonized countries have not regained access to their lands. Given this history, the Council believes that reparations are necessary. Many indigenous groups resisted a model of private "ownership," which went against philosophies that saw the land as living and agentic. They arrived at a philosophy of "Pathugappu," a caring stewardship of the land. While the Council has a long-term goal of creating a botanical commons, they began with a period where land and ideas could not be stolen.

3. We respect community ownership of their knowledge and heritage. We cannot appropriate this knowledge in the name of science or survival—doing so would only usher in a project of recolonization.

4. What is a successful species? By all accounts, bacteria have successfully inhabited earth since early life forms emerged on the planet. They have been so successful that life on earth has throughout been the "Age of Bacteria." For all the celebration of human intellect and superiority, we are galloping toward global annihilation.

5. Science has been critical to maintaining sex, gender, sexuality, nation, class, caste, and ability as biological categories.

6. With the exception of carnivorous plants and photosynthetic bacteria, algae, and fungi.

7. We worry that the term *sexual* is overdetermined, given more importance than necessary. The colonial version of the sexual (a racialized and binary view of sex/gender/sexuality that created a complementary and normative heterosexual mating system) has grounded biological thinking. To the chagrin of many of us, those that did not follow such a system were deemed *asexual* (a negation) or *vegetative* (passive zombies). We are troubled by both terms. Queer members of the Council have successfully incorporated new terminologies and vocabularies from global queer movements. We embrace the liberatory possibilities of these terms and embrace the "sexual" as a celebration of multiple possibilities. In embracing the term, we embrace the many queer transnational, transracial, crip, and indigenous imaginations and ways of being.

8. We use the term *photosynthetic organisms* (PSOs) because it includes bacteria, symbiotic fungi, algae, and animals. Thus we avoid the divisive logics of our inherited taxonomic systems.

9. We realize we work against a history where the idea of "green" was hijacked by failed projects of "carbon violence," where in the name of regreening, poor land-based communities were evacuated and marginalized for the sake of planetary survival. We deplore this history and have made sure that elites no longer control the process. See Kristen Lyons and Peter Westoby, "Carbon Markets and the New 'Carbon Violence': A Ugandan Study," *International Journal of African Renaissance Studies: Multi-, Inter-, and Transdisciplinarity* 9, no. 2 (2014): 77–94.

10. Evolution involves adaptations to environments. Where pollination or dispersal is difficult, inbreeding mechanisms and the ability to clone are critically important. We have been careful in our selections to accommodate the changing environmental conditions worldwide.

11. In "Animal Transsex," Myra Hird argues that "most plants are intersex, most fungi have multiple sexes, many species transsex, and bacteria completely defy notions of sexual difference, this means that the majority of living organisms on this planet would make little sense of the human classification of two sexes, and certainly less sense of a critique of transsex based upon a conceptual separation of nature and culture." Myra Hird, "Animal Transsex," in *Queering the Non/Human*, ed. Noreen Giffney, Myra J. Hird (London: Routledge, 2008), 237.

12. As Lewis Thomas argues: "My mitochondria comprise a very large portion of me. Looked at in this way, I could be taken for a very large, motile colony of respiring bacteria, operating a complex system of nuclei, microtubules, and neurons for the pleasure and sustenance of their families, and running, at the moment, a typewriter" (Thomas, *The Lives of Cells: Notes of a Biology Watcher* [New York: Viking, 1974], 72).

13. Margulis and Sagan, *Origins of Sex*.

14. Bacteria are key to life. Cyanobacteria, the early PSOs, brought oxygen to the atmosphere, enabling a diverse life on earth. Cell organelles like mitochondria and chloroplasts have their own DNA. Originally introduced by Constantin S. Merschkowsky, the "symbiogenesis theory" proposed the critical role of symbiosis in major evolutionary innovation. Popularized more recently by Lynn Margulis, this theory argues that eukaryotes emerged from endosymbiosis (an organism living within the body of another) and symbiogenesis (the merging of two separate organisms to form a single new organism). The fact that organelles like mitochondria and chloroplasts are similar in size to bacteria and have their own DNA is good evidence.

15. Bacteria, an early life form, are single-celled prokaryotes. They began as anaerobic creatures—that is, they did not need oxygen. Eventually, about 2.5 billion years ago, aerobic bacteria emerged, releasing oxygen as a byproduct of photosynthesis. Life is possible because of these bacteria. Their DNA is in a single genome within the cell. Bacterial biomass is greater than the biomass all life on earth combined, and in humans, bacterial cells outnumber our own. Bacteria are critical to life since they help with digestion. Bacteria are also the decomposers of the planet, helping in the cycling of nitrogen and carbon.

16. We are deeply suspicious of moves to "re-wild" spaces. The concept of "wilderness" places humans outside of nature. In the twenty-first century, there are efforts/movements to return to an imagined and mythic past. We reject all nostalgia of the colonial imagination. The world was never "wild" for all creatures—only for elite colonists. We reject this language of colonialism.

17. First, we are not recreating Noah's Ark, a foundational premise of savior science, where the focus is on saving individual species rather than sustaining diverse ecologies. Second, the classification of organisms imposes a certain "order." Both the morphologically grounded taxonomies of Linnaeus and the newer taxonomies of shared ancestry evidenced through molecular markers are artificial systems. They emerge from colonial attempts to impose systems of order, which unimaginatively reduce organisms to certain characteristics or genetic codes. They ignore the playful and wondrous world of biological innovations. No doubt these classification systems are sometimes useful, but to elevate them as "Truth" about evolved biologies of analogous or homologous traits seems undesirable. We reject such classification as the grounds to make decisions about the future.

18. In particular, we rely on the *Hortus* experiments across the globe that grew a stunning variety of plant communities. These studies yielded important insights on the changing climate and the adaptive abilities of organisms. The results of interspecies survival and success have helped guide choices across the globe.

19. We are suspicious of much of the Eurocentric and anthropomorphic language around ecology, with terms such as *native, alien, exotic, foreign, invasive, colonizers, pioneers*, and so on.

CHAPTER SEVEN | *Botanical Amnesia*

1. Crosby, *Ecological Imperialism.*

2. Mann, *1493*, 6.

3. Caroline Williams, "Pangaea, the Comeback," *New Scientist*, October 20, 2007.

4. For a more detailed discussion, see Subramaniam, *Ghost Stories.*

5. See Subramaniam, *Ghost Stories.*

6. Nancy Tomes, "The Making of a Germ Panic, Then and Now," *American Journal of Public Health* 90, no. 2 (February 2000): 191–99.

7. Draz, "On Gender Neutrality."

8. Wynter, "Unsettling the Coloniality," 260.

9. Mignolo, "Introduction."

10. Müller-Wille, "Linnaeus and the Four Corners of the World."

11. Wynter, "Unsettling the Coloniality."

12. See also Schiebinger, *The Mind Has No Sex*; Ordover, *American Eugenics*; Fausto-Sterling, "Gender, Race, and Nation"; Evelynn M. Hammonds and Rebecca M. Herzig, eds., *The Nature of Difference: Sciences of Race in the United States from Jefferson to Genomics* (Cambridge, MA: MIT Press, 2009); Donna J. Haraway. *Primate Visions: Gender, Race, and Nature in the World of Modern Science* (New York: Routledge, 2013); Rebecca M. Herzig, *Plucked* (New York: New York University Press, 2015).

13. Charles Elton, *The Ecology of Invasions by Animals and Plants* (Chicago: University of Chicago Press, 2000).

14. Matthew Hall, *The Imagination of Plants: A Book of Botanical Mythology* (New York: SUNY Press, 2019).

15. Mark Davis, *Invasion Biology* (New York: Oxford University Press, 2009).

16. Forest Service, US Department of Agriculture, "Invasive Species," https://www.fs.usda.gov/managing-land/invasive-species (accessed May 30, 2023).

17. Here I summarize key arguments from Subramaniam, *Ghost Stories*.

18. Mann, *1493*.

19. Anna Diamond, "America's First 'Food Spy' Traveled the World Hunting for Exotic Crops," *Smithsonian Magazine*, January 2018, https://www.smithsonianmag.com/arts-culture/smalltalk_fairchild-180967508/.

20. Juliet Lamb, "What If We Had All the Birds from Shakespeare in Central Park?," *JSTOR Daily*, June 9, 2016: https://daily.jstor.org/all-the-birds-from-shakespeare-in-central-park/.

21. Crosby, *Ecological Imperialism*.

22. Subramaniam, *Ghost Stories*.

23. Arturo Escobar, *Territories of Difference: Place, Movements, Life, Redes* (Durham, NC: Duke University Press, 2008.)

24. Laura A. Ogden, "The Beaver Diaspora: A Thought Experiment," *Environmental Humanities* 10, no. 1 (2018): 63–85.

25. Subramaniam, *Ghost Stories*, 122.

26. Cardozo and Subramaniam, "Assembling Asian America."

27. Jeannie N. Shinozuka, *Biotic Borders: Transpacific Plant and Insect Migration and the Rise of Anti-Asian Racism in America, 1890–1950* (Chicago: University of Chicago Press, 2022).

28. Alan Burdick, "The Truth about Invasive Species: How to Stop Worrying and Learn to Love Ecological Intruders," *Discover Magazine* 26, no. 5 (May 2005); Ian Frazier, "Fish Out of Water: The Asian Carp Invasion," *New Yorker* (2010): 66.

29. Clinton Crockett Peters, *Pandora's Garden: Kudzu, Cockroaches, and Other Misfits of Ecology* (Athens: University of Georgia Press, 2018).

30. Janet Marinelli and John M. Randall, *Invasive Plants: Weeds of the Global Garden* (Brooklyn: Brooklyn Botanic Garden, 1996).

31. Brockway, *Science and Colonial Expansion*.

32. Lorraine Daston and Katherine Park, *Wonders and the Order of Nature, 1150–1750* (Brooklyn: Zone Books, 2001).

33. Anthony R. Bean, "A New System for Determining Which Plant Species Are Indigenous in Australia," *Australian Systematic Botany* 20 (2007): 1–43; Matthew Chew and Andrew L. Hamilton, "The Rise and Fall of Biotic Nativeness: A Historical Perspective," in *Fifty Years of Invasion Ecology: The Legacy of Charles Elton*, ed. David M. Richardson (West Sussex: Wiley-Blackwell, 2011), 35–48; Hildegard Klein, comp., "Weeds, Alien Plants and Invasive Plants," PPRI Leaflet Series: Weeds Biocontrol, no. 1.1 (Pretoria: ARC-Plant Protection Research Institute, 2002); Petr Pyšek, David M. Richardson,

Marcel Rejmánek, Grady L. Webster, Mark Williamson, and Jan Kirschner, "Alien Plants in Checklists and Floras: Towards Better Communication between Taxonomists and Ecologists," *Taxon* 53 (2004): 131–43.

34. Scott Lauria Morgensen, "Theorising Gender, Sexuality and Settler Colonialism: An Introduction," *Settler Colonial Studies* 2, no. 2 (2012): 2–22.

35. Mark A. Davis, Ken Thompson, and J. Philip Grime, "Charles S. Elton and the Dissociation of Invasion Ecology from the Rest of Ecology,." *Diversity and Distributions* 7, no. 1-2 (2001): 97–102; Mark Davis, *Invasion Biology*; Mark Davis, "Let's Welcome a Variety of Voices to Invasion Biology," *Conservation Biology* 34, no. 6 (2020): 1329–1330; Robert I. Colautti and Hugh J. MacIsaac, "A Neutral Terminology to Define 'Invasive' Species," *Diversity and Distributions* 10, no. 2 (2004): 135–141.

36. Giovanni Vimercati, Anna F. Probert, Lara Volery, Ruben Bernardo-Madrid, Sandro Bertolino, Vanessa Céspedes, Franz Essl, Thomas Evans, Belinda Gallardo, Laure Gallien, Pablo González-Moreno, et al., "The EICAT+ Framework Enables Classification of Positive Impacts of Alien Taxa on Native Biodiversity," *PLoS biology* 20, no. 8 (2022), p.e3001729.

37. Subramaniam, *Ghost Stories*.

38. Richard J. Hobbs and David M. Richardson, "Invasion Ecology and Restoration Ecology: Parallel Evolution in Two Fields of Endeavour," in *Fifty Years of Invasion Ecology: The Legacy of Charles Elton*, ed. David M. Richardson, 61–70; Michelle Marvier, Peter Kareiva, and Michael Neubert, "Habitat Destruction, Fragmentation, and Disturbance Promote Invasion by Habitat Generalists in a Multispecies Metapopulation," *Risk Analysis* 24, no. 4 (2004): 869–878.

39. Richard Peet, *Unholy Trinity: The IMF, World Bank and WTO* (New York: Bloomsbury Publishing, 2009).

40. Patricio Javier Pereyra, "Revisiting the Use of the Invasive Species Concept: An Empirical Approach," *Austral Ecology* 41, no. 5 (2016): 519–528; Mark Davis, "Let's Welcome."

41. Valéry Loïc, Hervé Fritz, Jean-Claude Lefeuvre, and Daniel Simberloff, "Invasive Species Can Also Be Native . . . ," *Trends in Ecology & Evolution* 24, no. 11 (2009): 585.

42. Sonia Shah, "Native Species or Invasive? The Distinction Blurs as the World Warms," *Yale Environment* 360, January 14, 2020: https://e360.yale.edu/features/native-species-or-invasive-the-distinction-blurs-as-the-world-warms.

43. Menno Schilthuizen, *Darwin Comes to Town: How the Urban Jungle Drives Evolution* (New York: Picador, 2019); Laura F. Rodriguez, "Can Invasive Species Facilitate Native Species? Evidence of How, When, and Why These Impacts Occur," *Biological Invasions* 8, no. 4 (2006): 927–939; Jacques Tassin and Christian A. Kull, "Facing the Broader Dimensions of Biological Invasions," *Land Use Policy* 42 (January 2015): 165–169; Samuel Case, "Invasive Species Can Sometimes Help an Ecosystem," *Scientific American*, June 12, 2021.

44. Charles R. Warren, "Perspectives on the 'Alien' versus 'Native' Species Debate: A Critique of Concepts, Language and Practice," *Progress in Human Geography* 31 (4), 2007: 427–46; Mark Sagoff, "Who Is the Invader? Alien Species, Property Rights, and the Police

Power," *Social Philosophy and Policy* 26, no. 2 (2009): 26–52; Johan Hattingh, "Conceptual Clarify, Scientific Rigour and 'the Stories We Are': Engaging with Two Challenges to the Objectivity of Invasion Biology," in *Fifty Years of Invasion Ecology: The Legacy of Charles Elton*, ed. David M. Richardson, 359–75; Lesley Head, "The Social Dimensions of Invasive Plants," *Nature Plants* 3, no. 6 (2017): 1–7; Chew and Hamilton, "The Rise and Fall of Biotic Nativeness."

45. Marcus Hall, "Invasives, Aliens, and Labels Long Forgotten: Toward a Semiotics of Human-Mediated Species Movement," in *Humans Dispersals and Species Movement: From Prehistory to the Present*, ed. Nicole Boivin, Rémy Crassard, and Michael Petraglia (Cambridge, UK: Cambridge University Press, 2017): 430–453.

46. Hall, "Invasives, Aliens."

47. Lesley Head, "Decentring 1788: Beyond Biotic Nativeness," *Geographical Research* 50, no. 2 (2012): 166–178.

48. Benjamin D. Hoffmann and Franck Courchamp, "Biological Invasions and Natural Colonisations: Are They That Different?," *NeoBiota* 29 (2016): 1–14;

49. Tomaz Mastnak, Julia Elyachar, and Tom Boellstorff, "Botanical Decolonization: Rethinking Native Plants," *Environment and Planning D: Society and Space* 32, no. 2 (2014): 363–380.

50. Mastnak, Elyachar, and Boellstorff, "Botanical Decolonization."

51. Lesley Head and Pat Muir, "Nativeness, Invasiveness, and Nation in Australian Plants," *Geographical Review* 94, no. 2 (2004): 199–217.

52. Jozef Keulartz and Cor van der Weele, "Framing and Reframing in Invasion Biology," *Configurations* 16, no. 1 (2008): 93–115.

53. William Jordan, "The Nazi Connection," *Restoration & Management Notes* 12 (1994): 113; Ned Hettinger, "Exotic Species, Naturalisation, and Biological Nativism," *Environmental Values* 10 (2001): 217.

54. Keulartz and Weele, "Framing and Reframing in Invasion Biology."

55. Shashi Tharoor, *Inglorious Empire: What British Did to India* (London: Penguin, 2017).

56. Simpson, "Temporary Spaces."

57. William O'Brien, "Exotic Invasions, Nativism, and Ecological Restoration: On the Persistence of a Contentious Debate," *Ethics, Place & Environment* 9 (2006): 73.

58. Head, "Decentring 1788."

59. Pyšek et al., "Alien Plants in Checklists and Floras."

60. TallBear, review of *All Our Relations*.

61. Jessica Cattelino, "Loving the Native," in *The Routledge Companion to the Environmental Humanities*, ed. Ursula K. Heise, Jon Christensen, and Michelle Niemann (London: Routledge, Taylor and Francis Group, 2017).

62. Nicholas J. Reo and Laura A. Ogden. 2018, "Anishnaabe Aki: An Indigenous Perspective on the Global Threat of Invasive Species," *Sustainability Science* 13, no. 5 (2018): 1443–1452.

63. Jessica Hernandez, *Fresh Banana Leaves: Healing Indigenous Landscapes through Indigenous Science* (Berkeley, CA: North Atlantic Books, 2022).

64. Jessica Hernandez, "Invasive Species as a Metaphor for Colonization," *Rewilding Magazine*, June 16, 2022.

65. TallBear, "Caretaking Relations."
66. Reo and Ogden, "Anishnaabe Aki."
67. Head, "Decentring 1788."

CHAPTER EIGHT | *Like a Tumbleweed in Eden*

1. John Leland, *Aliens in the Backyard: Plant and Animal Imports into America*, (Columbia: University of South Carolina Press, 2005). Recent genetic studies update our understanding of origin of horses in North America. See William Timothy Treal Taylor, Pablo Librado, Mila Hunska Tašunke Icu (Chief Joseph American Horse), Carlton Shield Chief Gover, Jimmy Arterberry, Anpeta Luta Wiŋ (Antonia Loretta Afraid of Bear–Cook), Akil Nujipi (Harold Left Heron), et al., "Early Dispersal of Domestic Horses into the Great Plains and Northern Rockies," *Science* 379, no. 6639 (2023): 1316–1323.

2. Zadie Smith, *White Teeth* (New York: Vintage, 2000).

3. Leland, *Aliens in the Backyard.*

4. Cardozo and Subramaniam, "Assembling Asian America."

5. Paul Gilroy, "Diaspora," *Paragraph* 17, no. 3, Keywords (November 1994): 207–212; B. H. Edwards, "The Uses of Diaspora," *Social Text* 19, no. 1 (2001): 45–73. Thanks to Samantha Pinto for her insight.

6. Gilroy, "Diaspora."

7. Jan Surman, Katalin Straner, and Peter Haslinger, "Introduction: Nomadic Concepts: Biological Concepts and Their Careers beyond Biology," *Contributions to the History of Concepts* 9, no. 2 (2014): 1–17.

8. For more discussion of diasporas and empire, see Gary B. Magee, *Empire and Globalization: Networks of People, Goods and Capital in the British World, c. 18550–1914* (New York: Cambridge University Press, 2010).

9. James C. Scott, *Seeing Like a State: How Certain Schemes to Improve the Human Condition Have Failed* (New Haven, CT: Yale University Press, 1999).

10. I should note that Achebe further reflects on this proverb later in his work in *Home and Exile* (New York: Oxford University Press, 2000).

11. In particular, see Subramaniam, *Ghost Stories.*

12. Bruno Latour, *We Have Never Been Modern* (Cambridge, MA: Harvard University Press, 1993).

13. See Crosby, *Ecological Imperialism*; Grove, *Green Imperialism*; William Beinart and Lotte Hughes, *Environment and Empire* (Oxford: Oxford University Press, 2007); Huggan and Tiffin, *Postcolonial Ecocriticism*; Elizabeth M. DeLoughrey and George B. Handley, eds., *Postcolonial Ecologies: Literatures of the Environment* (Oxford: Oxford University Press, 2011).

14. Grove, *Green Imperialism*; Schiebinger and Swan, *Colonial Botany.*

15. See Crosby, *Ecological Imperialism*; Jeffrey A. McNeely, *The Great Reshuffling: Human Dimensions of Invasive Alien Species* (Gland, Switzerland: IUCN, 2001); Warren, "Perspectives"; Heather Weiner, "Congress Threatens Wild Immigrants," *Earth Island Journal* 11, no. 4 (1996).

16. Crosby, *Ecological Imperialism*, 7.

17. Brockway, *Science and Colonial Expansion.*

18. Grove, *Green Imperialism.*

19. DeLoughrey and Handley, *Postcolonial Ecologies.*

20. Bernd Lenzner, Guillaume Latombe, Anna Schertler, Hanno Seebens, Qiang Yang, Marten Winter, Patrick Weigelt, et al., "Naturalized Alien Floras Still Carry the Legacy of European Colonialism," *Nature, Ecology & Evolution* 6, no. 11 (2022): 1723–1732.

21. Subramaniam, *Ghost Stories.*

22. Adam Rone, "Nature Wars, Culture Wars: Immigration and Environmental Reform in the Progressive Era," *Environmental History* 13, no. 3 (July 2008): 432–453.

23. Foucault, *The Order of Things.*

24. Grove, *Green Imperialism.*

25. Cited in DeLoughrey, and Handley, *Postcolonial Ecologies,* 29.

26. Schiebinger, *Nature's Body.*

27. Markowitz, "Pelvic Politics."

28. Markowitz, "Pelvic Politics."

29. Markowitz, "Pelvic Politics."

30. DeLoughrey and Handley, *Postcolonial Ecologies.*

31. For a longer discussion, see Subramaniam, *Ghost Stories.*

32. Chew and Hamilton, "The Rise and Fall of Biotic Nativeness."

33. Mark A. Davis, Matthew K. Chew, Richard J. Hobbs, Ariel E. Lugo, John J. Ewel, Geerat J. Vermeij, James H. Brown, et al., "Don't Judge Species on Their Origins," *Nature* 474, no. 7350 (2011): 153–154.

34. Philip Pauly, "The Beauty and Menace of the Japanese Cherry Trees: Conflicting Visions of American Ecological Independence," *Isis* 87, no 1 (1996): 51–73.

35. Latour, *We Have Never Been Modern.*

36. For a longer discussion and citations for the various media claims, see Subramaniam, *Ghost Stories.*

37. Tomes, "The Making of a Germ Panic."

38. Tomes, "The Making of a Germ Panic."

39. Mark Robichaux, "Alien Invasion: Plague of Asian Eels Highlights Damage from Foreign Species," *Wall Street Journal*, September 27, 2000, 12A.

40. Mark Cheater, "Alien Invasion," *Nature Conservancy* 42, no. 5 (1992): 24–29.

41. Joseph B. Verrengia, "Some Species Aren't Welcome," *ABC News*, September 27, 2000.

42. Christopher Bright, "Invasive Species: Pathogens of Globalization," *Foreign Policy* 116 (1999): 50–60, 62–64; quote is from 51.

43. Mark Davis et al., "Don't Judge Species on Their Origins"; Hugh Raffles, "Mother Nature's Melting Pot," *New York Times*, April 2, 2011.

44. Mark Davis et al., "Don't Judge Species on Their Origins."

45. Martin A. Schlaepfer, Dov F. Sax, and Julian D. Olden, "The Potential Conservation Value of Non-Native Species," *Conservation Biology* 25, no. 3 (2011): 428–437.

46. Patrick C. Tobin, "Managing Invasive Species," *F1000Research* 7 (2018).

47. Mark Davis et al., "Don't Judge Species on Their Origins."

48. Raffles, "Mother Nature's Melting Pot."

49. Robert Assarello, *Queer Environmentality: Ecology, Evolution, and Sexuality in American Literature* (Burlington, VT: Ashgate Publishing, 2012).

50. Gwendolyn Brooks, "To the Diaspora," 1981, Modern American Poetry, https://www.modernamericanpoetry.org.

INTERLUDE | *Love the Dandelion*

1. Rivera Sun, *The Dandelion Insurrection: Love and Revolution* (El Prado, NM: Rising Sun Press Works, 2013).

INTERLUDE | *A Cosmopolitan Botany*

1. Ajit Mondal and Jayanta Mete, "Tagore's Santiniketan School: A Retrospective View," *Bangladesh Education Journal* 13, no. 2 (2014): 49–57.

2. Evelyn Ramos, "Rabindranath Tagore: An Indian Polymath," Rubin Museum of Art Blog, August 3, 2018, https://rubinmuseum.org/blog/rabindranath-tagore-an-indian-polymath.

3. Rabindranath Tagore, "My School" (lecture delivered in America and published in *Personality* London: MacMillan, 1933), http://home.iitk.ac.in/~amman/soc748/tagore_myschool.html (accessed September 2, 2022).

4. Quoted in Mondal and Mete, "Tagore's Santiniketan School."

5. Melanie R. Clark, "Design without Borders: Universalism in the Architecture of Rabindranath Tagore's 'World Nest' at Santiniketan" (master's thesis, Brigham Young University, 2020), https://www.proquest.com/docview/2430937698?fromopenview=true&pq-origsite=gscholar.

6. Clark, "Design without Borders."

7. Clark, "Design without Borders."

8. Ratan Lal Basu, "Viewpoint: The Eco-Ethical Views of Tagore and Amartya Sen," *Culture Mandala: Bulletin of the Centre for East-West Cultural and Economic Studies* 8, no. 2 (December 2009): 56–61.

9. Quoted in Argha Banerjee, "Tagore and the Eco-Friendly Campus," *The Statesman*, February 9, 2017, https://www.thestatesman.com/opinion/tagore-and-the-eco-friendly-campus-1486677320.html (accessed November 16, 2023).

10. UNESCO World Heritage Convention, 2020, "Santiniketan": https://whc.unesco.org/en/tentativelists/5495/ (accessed September 2, 2022).

11. Clark, "Design without Borders."

12. Sumit Bhattacharjee, "Tagore's Vision of an Institution," *The Hindu*, March 26, 2012.

13. A. K. Rai, "Rabindranath's Vision of Freedom and Openness Has Been Vandalized in Santiniketan," *Kalimpong News* April 30, 2021.

14. Lily Roy, "Tagore's Views on the Religion of the Forest and Relationship of Man and Nature," *International Journal of Research and Analyti Views* 15, no. 3 (2018).

15. Lily Roy, "Tagore's Views."

16. Lily Roy, "Tagore's Views."

CHAPTER NINE | *Vegetal Sublimations*

1. Sophia Roosth and Astrid Schraeder, eds., "Feminist Theory Out of Science," special issue, *Differences* 23, no. 3 (2012).

2. Audre Lorde, "Age, Race, Class and Sex: Women Redefining Difference," in *Sister Outsider* (Crossing Press, 1984).

3. Karen Cardozo and Banu Subramaniam, "Assembling Asian/American Naturecultures: Orientalism and Invited Invasions," *Journal of Asian American Studies* 16, no. 1 (2013).

4. Many thanks to Jennifer Nash for this insight. Patricia J. Williams, *The Alchemy of Race and Rights* (Cambridge, MA: Harvard University Press, 1991).

5. Quoted in Jhala, "At a Cambridge University College."

6. TallBear, "Caretaking Relations."

7. Chela Sandoval, *Methodology of the Oppressed* (Minneapolis: University of Minnesota Press, 2000).

8. Luz Calvo, review of *Methodology of the Oppressed* by Chela Sandoval and *Decolonizing Methodologies: Research and Indigenous Peoples* by Linda Tuhiwai Smith, *Signs* 29, no. 1 (Autumn 2003), 254–257; quote is from 254. See also Linda Smith, *Decolonizing Methodologies*; Sandoval, *Methodology of the Oppressed.*

9. Philip, *Civilizing Natures*; Raj, *Relocating Modern Science*; Prasad, *Imperial Technoscience*; Mukharji, *Doctoring Traditions.*

10. Simpson, "Not Murdered and Not Missing."

11. Lorde, "Age, Race, Class and Sex."

12. Sumana Roy, *How I Became a Tree.*

13. Eve Tuck, "Suspending Damage: A Letter to Communities," *Harvard Educational Review* 79, no. 3 (2009): 409–428.

14. Maria Puig de La Bellacasa, *Matters of Care: Speculative Ethics in More than Human Worlds* (Minneapolis: University of Minnesota Press, 2017); TallBear, "Caretaking Relations"; Marisol De la Cadena, *Earth Beings: Ecologies of Practice across Andean Worlds* (Durham, NC: Duke University Press, 2015).

15. Chen, *Animacies*; Chanda Prescod-Weinstein, "Making Black Women Scientists under White Empiricism: The Racialization of Epistemology in Physics,: *Signs: Journal of Women in Culture and Society* 45, no. 2 (2020): 421–447; Chanda Prescod-Weinstein, *The Disordered Cosmos: A Journey into Dark Matter, Spacetime, and Dreams Deferred* (London: Hachette UK, 2021);

16. Keller, *A Feeling for the Organism*; Deboleena Roy, *Molecular Feminisms: Biology, Becomings, and Life in the Lab* (Seattle: University of Washington Press, 2018); Karen Barad, *Meeting the Universe Halfway: Quantum Physics and the Entanglement of Matter and Meaning* (Durham, NC: Duke University Press, 2007); Natasha Myers, "Ungrid-able Ecologies: Decolonizing the Ecological Sensorium in a 10,000 Year-Old Naturalcultural Happening," *Catalyst* 3, no. 2 (2017): 1–24.

17. Eduardo Kohn, *How Forests Think: Towards an Anthropology Beyond the Human* (Durham, NC: Duke University Press, 2013); Marsha Weisinger, *Dreaming of Sheep in Navajo Country* (Seattle: University of Washington Press, 2011); Haraway, *Staying with the Trouble*; Kimmerer, *Braiding Sweetgrass.*

18. Carla Hustak and Natasha Myers, "Involutionary Momentum: Affective Ecologies and the Sciences of Plant/Insect Encounters," *differences* 23, no. 3 (2012): 74–118; McKittrick, *Dear Science*; Anna Lowenhaupt Tsing, *The Mushroom at the End of the World: On the Possibility of Life in Capitalist Ruins* (Princeton, NJ: Princeton University Press, 2015); Anna L. Tsing, "Unruly Edges: Mushrooms as Companion Species: For Donna Haraway," *Environmental Humanities* 1, no. 1 (2012): 141–154.

19. Liboiron, *Pollution Is Colonialism*; Tuck and Wang, "Decolonization Is Not a Metaphor."

20. Catriona Mortimer-Sandilands and Bruce Erickson, *Queer Ecologies: Sex, Nature, Politics, Desire* (Bloomington: Indiana University Press, 2010); Willey, *Undoing Monogamy*; Wölfle Hazard, *Underflows.*

21. Kristina Lyons, *Vital Decomposition: Soil Practitioners and Life Politics* (Durham, NC: Duke University Press, 2020); Dimitris Papadopoulos, María Puig de La Bellacasa, and Natasha Myers, eds., *Reactivating Elements: Chemistry, Ecology, Practice* (Durham, NC: Duke University Press, 2021).

22. Alexis Shotwell, *Against Purity: Living Ethically in Compromised Times* (Minneapolis: University of Minnesota Press, 2016); Michelle Murphy, "Alterlife and Decolonial Chemical Relations," *Cultural Anthropology* 32 (2017), 497; Gayatri Spivak, "Planetarity," *Paragraph* 38, no. 2 (2015): 290–292;

23. Juno Salazar Parreñas, *Decolonizing Extinction: The Work of Care in Orangutan Rehabilitation* (Durham, NC: Duke University Press, 2018); Chacko, "Stringing, Reconnecting, and Breaking"; Xan Chacko, "Digging Up Colonial Roots: The Less-Known Origins of the Millennium Seed Bank Partnership," *Catalyst: Feminism, Theory, and Technoscience* 5, no. 2 (2019): 1–9; Linda Smith, *Decolonizing Methodologies.*

24. Donna Haraway, "Situated Knowledges: The Science Question in Feminism and the Privilege of Partial Perspective," *Feminist Studies* 14, no. 3 (1988): 575–599.

25. Robert N. Proctor and Londa Schiebinger, *Agnotology: The Making and Unmaking of Ignorance* (Stanford, CA: Stanford University Press, 2008).

26. Walter Rodney, *How Europe Underdeveloped Africa* (New York: Verso Books, 2018).

27. Kincaid, *My Garden (Book)*, 115.

28. Julietta Singh, *Unthinking Mastery: Dehumanism and Decolonial Entanglements* (Durham, NC: Duke University Press, 2018), 158.

29. Mohsin Hamid, "Mohsin Hamid on the Rise of Nationalism: 'In the Land of the Pure, No One Is Pure Enough,'" *The Guardian*, January 27, 2018.

CHAPTER TEN | *Dreams of a Lively Planet*

1. Noble, *A World without Women.*

2. Amitav Ghosh, "The Nutmeg's Curse," in *The Nutmeg's Curse* (Chicago: University of Chicago Press, 2021), 129.

3. George Lipsitz, "The White Possessive and Whiteness Studies," *Kalfou* 6, no. 1 (2019): 42–51.

4. Samir Amin, *Eurocentrism* (New York: NYU Press, 1989).

5. Faye V. Harrison, ed., *Decolonizing Anthropology: Moving toward an Anthropology for Liberation*, 3rd ed. (Arlington, VA: American Anthropological Association, 2010).

6. Tania Pérez-Bustos, "A Word of Caution toward Homogenous Appropriations of Decolonial Thinking in STS," *Catalyst: Feminism, Theory, Technoscience* 3, no. 1 (2017).

7. Tuck and Yang, "Decolonization Is Not a Metaphor," 3.

8. Thanks to Lukas Rieppel for pushing me to better articulate the material stakes of decolonization.

INTERLUDE | *The Memory Gardens*

1. In Octavia Butler's memorable words.

INTERLUDE | *Abolitionist Futures*

1. Thanks to Karen Carodozo for this insight.

INDEX

ableism, 9, 12, 174; in the academy, 67; anti-, 246, 250; in writings about plants, 22, 149
academic disciplines. *See* disciplines, academic; interdisciplinarity
Achebe, Chinua, 199, 202
Achuthan, Itty, 110, 112
adisciplinary sciences, 227–228; cartographies for, 233–235; methodologies for the pressed in, 228–232
Against Purity (Shotwell), 77
Age of Exploration, 138–139
agnotology, 234
agriculture, 20, 32, 92, 161, 188; climate change effects on, 228; colonialism and European, 203; indigenous, 162, 164; industrialized, 162–165; modern, 162–163; North American, 210. *See also* invasion biology
alternative medicine, 113–118
Ambedkar, B. R., 119
American Acclimatization Society, 188
American Civilian Conservation Corps, 189
American Ornithological Society, 97
American West, 199
Amin, Samir, 237–238
amnesia, botanical. *See* botanical amnesia
androdioecious plants, 142
androecious plants, 142
androgynomonoecious plants, 142
andromonoecious plants, 142
anisogamy, 145, 177
Anthropocene, 52, 194
anthropocentrism, 147, 191, 206
anthropomorphism, 148–149, 230; Linnaeus's, 2, 144
anthropos, ix, 52, 206
antiableism, 246, 250
anti-binary, 246
anti-biodiversity, 245–246
anticasteism, 119, 246. *See also* caste
anticolonialism: avoidance of term, 53; and future, 246; impact of learning from, 26; invasive species and, 194; subverted by Hindu nationalists, 53, 57, 121, 249; Tagore and, 223; using insights from, 228, 233; and Vedic sciences, 73
anti-disciplines, 246
anti–Edenic science, 194, 243–246
anti-individual, 172, 176, 245
antiracism: and future, 246; using insights from, 228, 233
anti-universalism, 244–245
Anzaldúa, Gloria, 152, 155, 227
Aristotle, 56
Asher, Kiran, 155
Attenborough, David, 34–35
Audubon Society, 97
Ayurveda, 111–112, 115, 116–117; and TKDL, 115, 118

Baartman, Sara, 156
bacteria, 292nn14–15
Bacteria Project, 173
Bagemihl, Bruce, 133, 168
Banyan Project, 171
Barad, Karan, 231
Barcode of Life (BoL), 95
Bartlett, Madelaine, 144
Bateman, A. J., 146, 149, 150–151; critiques of, 146–147, 150–151
Bauhin, Gaspard, 81
Beamtimes and Lifetimes (Traweek), 39
Beanion Project, 172

Beck, Martha, 4
Beloved (Morrison), 8
Benally, Klee, 14
Bentham, George, 27
Berlant, Lauren, 166
Bessey, Charles Edwin, 27
Between the World and Me (Coates), 225
Bhat, Appu, 110
Bhat, Ranga, 110
Bil, Geoff, 88
binary, 4, 11, 18–19, 55, 232, 259n35; anti-, 246; diaspora and, 201; false separation of nature and culture, 30–31, 45–46; geographic, 206–207; native/alien, 192, 194, 206–207; Queer Vegennials, 170–171, 176–178; sexual, 136–137, 141–144, 152, 158–159
biodiversity: anti-, 245–246; and colonialism, 122, 129; in India, 106; and indigenous knowledge systems, 7; management of, 95; in nature vs. herbaria, 17
biogeography: botanical amnesia and, 186–191; colonial hauntings in, 191–193; Columbian Exchange and, 184; cultivating living relations and, 197–198; idea of nativeness in, 193–196; native vs. foreign plants and, 185; race and coloniality of power and, 186
Biological Exuberance (Bagemihl), 133, 168
biopiracy, 9, 84, 113–114, 122, 227
Bitter Roots (Osseo-Asare), 96
"Black Feminist Hauntology" (Saleh-Hanna), 8
Black Lives Matter (BLM), 166
Blackness, 154
Bleier, Ruth, 146
Bondestam, Maja, 143
Bonilla, Yarimar, 58
botanical amnesia, 9, 18, 152, 186–191, 197–198, 227, 233, 236
botanical cosmopolitanism, 31–32, 93, 194, 196
botanical infrastructure, 80, 97, 184
botany, 1–2; abolitionist science and, 247–252; adisciplinary reimagining of, 227–235; bringing order to plant worlds, 79–80; coloniality of (*see* coloniality); ethno-, 21; feminist histories of, 2–3; fieldwork in, 40–41; history of, 2–4, 27–28, 81–84; as imperial archive, 94–98; of India, 29–30; infrastructures of naming, nomenclatures, and classifications in, 92–94; interdisciplinarity in, 6–7, 8, 47–48; key conceptual terrains of, 5–10; nomenclature in, 80, 81–84; rememory in, 9, 152; renaming of plants in, 21–22; scientization of, 18, 137; stories we tell about, 50–51; study of, in India, 28–29, 32, 34, 36; subfields of, 14–15; sublime, 26–27; Subramaniam's love of, 25–26, 29–32, 36–38; taxonomy and systematics of, 80, 81–84; television shows and, 34–35; tree of life and, 90–91, 100–102. *See also* science
Bowker, Geoffrey C., 92, 95
Braiding Sweetgrass (Kimmerer), 41
Brand, Dionne, 153
Briggs, Laura, 161
British Museum, 139, 205
Brooks, Gwendolyn, 212
Browne, Janet, 137, 139, 205
Bryson, Bill, 90
Buddhism, 119, 219
Butler, Octavia, 152, 232

Caesalpino, Andrea, 81
caretaking relations, 198
"Caretaking Relations, Not American Dreaming" (TallBear), 49
caste, 37, 52, 72, 234, 250; lower/marginalized, 65–67, 110–111; modern systems of, 247–248; politics of, 119–121; social movements against discrimination based on, 112–113 (*see*

also anticasteism); upper, 37, 57, 109–111, 118, 122–123, 220, 221, 223
Cattelino, Jessica, 197
Chacko, Xan, 164
Chen, Mel, 156, 231
Chinese Exclusion Act, US, 207
Chirp-Net, 60–64, 125–126, 170, 171, 213, 239
Christian culture, 86–87, 99, 140, 175, 186; clerical tradition, 36
Cibo, Gherardo, 50
cis-heterosexual reproduction, 149–150
cladistics, 91
classification, 92–94, 137, 139, 293n17; lumpers and splitters, 79–80. *See also* naming, plant
Cleetus, Burton, 110
Clements, Frederick, 27
climate change, 35, 64, 84, 153, 192–193, 210, 214, 237, 242; empire and, 228–229; queer botany and, 174; urgency of attention on, 164, 166–167
clonal reproduction, 145
Coates, Ta-Nehisi, 225
colonialism, 2, 3–4, 53; abolitionist futures and, 247–252; Age of Exploration and, 138–139; biogeography and, 191–193; botany as central to, 4–5, 86–89, 150–151, 203; Christian culture and, 86–87, 99, 140; controlling biology and reproduction in, 160–165; counter-, 55; decolonization and, 10, 53, 55–58, 236–238; embranglements and, 7–8; extractive economy and, 46, 185; gender and, 155–156; as genocide, ecocide, and epistemicide, 20, 52; geographies and histories of, 53–54; history of botany and, 27–28; in India, 32–33; Indigenous systems and peoples and, 83–84; invasion biology and, 187–189; as knitting the world together, 236–237; legacies of, 55; lives and afterlives of, 118–120; native knowledge and, 45–46; plant, 86–89, 161–165; post-, 57–58, 120–121, 122–123, 202; racial capitalism in, 11, 163–164; rememory and, 8; repression by the state and, 54–55; spices and, 103–104; story of empire and, 50–51; terraforming and, 58–59; utility and, 83
coloniality, 7, 47; Christian culture and, 99; defining, 54, 156–157; diversity and women in science and, 65–73; of gender, 51–52, 68–70, 155–157; of nomenclature, 104, 120–122; of power, 68–73, 98–99, 186, 237; of race, 51–52, 186 (*see also* race); of science, 51–54, 70–72
colonial logics, 162–165
colonial violence, 19, 152
Columbian Exchange, The (Crosby), 184
Columbus, Christopher, 83, 87, 103
conservationists, 44
Convention on Biological Diversity, 114
Cosmos (TV show), 34
countercolonialism, 55
Covid-19 pandemic, 184
crip theory, 12
crip time, 12–13
Cronon, William, 43, 44, 58
Crosby, Alfred, 184, 203–204
Cutting for Stone (Verghese), 25

da Gama, Vasco, 104
damage-centered research, 230–231
Dancing at the Edge of the World (Le Guin), 236
Darwin, Charles, 27, 34, 35, 90, 143, 146, 168–169; on sexual selection, 149, 151; work with orchids, 146–147, 152
Darwin, Erasmus, 139
Davis, Mark, 210
Dear Science (McKittrick), 127, 129–130
Death of a Discipline (Spivak), 202
Death of Nature (Merchant), 45
de Candolle, Alphonse, 95

decolonization, 10, 53, 55–58, 233–238; imperatives of, 237–238; and its (dis) contents, 54–58; as ongoing process, 20, 236–237; recolonization and, 57; reflection of ongoing colonial practices in, 95–96
de la Bellacasa, 231
de la Cadena, Marisol, 231
DeLoughrey, Elizabeth, 206
development, 44, 45, 79, 99; colonialism and, 118–119, 234; logics of, 189; over-, 192–193, 210, 228; sustainable, 221; under-, 234
diaspores and diasporas, 201–202, 207; politics of naming and, 210
differential consciousness, 230
disability studies, 11–14, 123
disciplines, academic: as centering the west and whiteness, 21; colonialism and coloniality in, 7, 9, 28–29, 50–51, 99, 227–228; narrowness of, 6, 47–48, 202, 234; reshaping of, 54, 203; sexuality and, 158; shaping of, 14, 154; terminology from, 15; unlearning narratives of, 10. *See also* interdisciplinarity
Divakaruni, Chitra Banerjee, 103
diversity and women in science, 65–73
DNA: and evolution, 90; movement and mixing of, 101, 167; taxonomy and, 125; technologies for studying, 86, 100. *See also* genetics
Doctrine of Signatures, 82
Doolittle, W. F., 101

Earth BioGenome Project, 95
ecocide ecosystems, 20, 52
Ecological Imperialism (Crosby), 203
ecology, 15, 163, 193–194, 203, 231; culture and, 210; Edenic science and, 245; explosion, 186; postcolonial, 206; restoration, 52, 202, 230
Ecology of Invasions by Animals and Plants, The (Elton), 186–187
Edenic ecologies, 88, 204, 212
Ehrenreich, Ben, 58
Elton, Charles, 186–187
embranglements, 7–8
empire: language of, 205–212; sensing like, 202–205
empirical knowledge, 88, 112, 144–145
empiricism, 231
enculturation, 41
Encyclopedia of Life (EoL), 95
Endersby, Jim, 85
Erickson, Bruce, 231
Escobar, Arturo, 189
ethnobotany, 21
ethnoscience, 6
eugenics, 11, 127, 160, 175, 233; laws and, 11; science of, 243; scripts of, 160, 164
evolutionary sciences, 89–91; sexual biology and, 158. *See also* genetics; natural selection
Evolution's Rainbow (Roughgarden), 168
extinction, 18, 95, 174, 210, 213, 239, 242
extractive economy, 46, 185

Fairchild, David, 188
Fanon, Frantz, 49, 55
Fausto-Sterling, Anne, 145, 151
feminist scholarship, 52; in botany, 2–3; on culture of no culture, 38, 39–40; feminist STS and, 38; on race, sexuality, and reproduction, 158–160; on science, 202
fertilization, 144, 145; self-, 148–149
Finch, Bill, 143–144
fitness, 145, 149
folk taxonomies, 81–84
Foster, Laura, 96
Foucault, Michel, 82, 98
Fournier, Marian, 108
Freedom from the Known (Krishnamurti), 77
futures: abolitionist, 247–252; accessible,

13–14; agricultural, 163; anti-Eden and, 243–246; botanical, 4, 14; colonialism, decolonization, and, 49, 238; evolutionary, 191; homeland idea and, 200; movement toward just, 8, 10, 14, 15, 224, 252; movement toward livable, 61–64, 178; nationalism and, 242; naturecultural, 19; rememory and, 9, 22

Gannett, Lisa, 101
gender: coloniality of, 155–157; kitchen spaces and, 105–106; regendering, 165; tied to race, 100–101, 139; and women in science, 65–73
Genera Plantarum (Jussieu), 85, 139
genetic parasitism, 145
genetics, 15, 36, 91, 101. *See also* DNA
genetic variation, 100, 145
genocide, 20, 52, 54
George, Sam, 2
Ghosh, Amitav, 104, 237
Ghostly Matters (Gordon), 8
Ghost Stories for Darwin (Subramaniam), 8, 22, 47, 66, 127, 159, 233
Gledhill, David, 92–93
Glissant, Edouard, 7
Gordon, Avery, 8
Gould, Stephen Jay, 134
Gowaty, Patricia, 283–284n63
Gray, Asa, 27
great chain of being, 55, 56, 206
Grew, Nehemiah, 136
Grosz, Liz, 152
Grove, Richard, 109–110, 204
Gumbs, Alexis Pauline, 156
gynodioecious plants, 142
gynoecious plants, 142
gynomonoecious plants, 142

Haeckel, Ernt, 90
Hamid, Mohsin, 235
Hammonds, Evelynn, 156
Handley, George, 206
Haraway, Donna, 6, 152, 163, 231
Harjo, Joy, 105
Harrison, Faye, 238
Hartman, Saidiya, 23, 160
Henslow, John, 207
herbaria, 17, 20, 50, 72, 129, 224; globalization of, 122; herbarium sheets and, 83, 108; plant identification and, 96
hermaphroditic plants, 141–143, 145
Hernandez, Jessica, 8
Herzig, Rebecca, 152
Hindu nationalism, 57–58, 78, 111, 117, 121, 224
Hindu tradition, 30
Hird, Myra, 152
Holiday, Billie, 154
Holy Science (Subramaniam), 47
Hooker, Joseph, 27
Hortus Malabaricus (Rheede), 21, 104, 106–113, 118, 120–121
How Europe Underdeveloped Africa (Rodney), 234
How to Queer Ecology (Johnson), 25
Hrdy, Sarah, 146
Human Genome Diversity Project, 100–101
Humboldt, Alexander, 88
Hurston, Zora Neale, 133
Hustak, Carla, 152, 231
Huxley, Julian, 100, 101

Ibe, Carol, 72
immigrants, 184, 187, 190–191, 204–205, 208–209, 211
immigration, 184–185, 190, 192, 205; germ panics surrounding, 208; politics of contemporary, 207, 209–211; rhizomatic networks and, 201
imperialism, 50–51, 87, 95; imperial archive, 94–98; imperial globality, 189
Imperial Leather (A. McClintock), 100

India, 77–78; British colonialism in, 32–33, 110, 119, 122; independence movement in, 119; kitchen spaces in, 105–106; language in, 119–120; mythology of trees in, 153–154; postcolonial education in, 28–29, 32, 34, 36; Subramaniam's childhood kitchen table in, 105–106; Tagore on, 218–224; television in, 34; Traditional Knowledge Database Library (TKDL) in, 113–118. *See also* Hindu nationalism; South Asia
indigenization, 56–57
Indigenous knowledge, 45–46, 57, 83–84, 291nn1–3; biogeography and, 197–198; local knowledges and, 21, 45–46, 117–118; Traditional Knowledge Database Library (TKDL) and, 113–118
Indigenous movements, 166
interdisciplinarity, 6–7, 8, 47–48
International Botanical Congress, 92–93
International Code of Botanical Nomenclature, 94, 97, 139
International Code of Nomenclature for Algae, Fungi, and Plants, 80
International Monetary Fund, 193
invasion biology, 16, 18–19, 184, 186–191; panic in Waisley over, 213–217; politics of naming and, 209–211; politics of purity and, 52; rhetoric of, 190, 192, 208, 218. *See also* nativeness

Johnson, Alex Carr, 25
Jones, William, 98, 122
Jussieu, A. L. de, 85, 86

Keller, Evelyn Fox, 40, 41, 231
Kelley, Robin, 238
Keulartz, Jozef, 195
Kew Gardens, 46, 139, 188, 205
Kimmerer, Robin Wall, 7, 8, 32–33, 41, 42, 58–59, 162
Kincaid, Jamaica, 33, 139–140, 232, 234–235
King, Amy, 86
Kinsman, Sharon, 167–168
kitchen spaces, 105–106, 123
Kitchen Table: Women of Color Press, 105–106
knowledge-making practices, 45–46. *See also* empirical knowledge; Indigenous knowledge; monoculture of knowledge; Traditional Ecological Knowledge; universal knowledge; Vedic knowledge
Koerner, Lisbet, 88
Kohn, Eduardo, 231
Krishnamurti, Jiddu, 77
Kroerner, Lisbeth, 87
kudzu, 185, 189–190, 200, 216

LaFleur, Greta, 155
Lamarck, Jean-Baptiste, 27
language of empire, 205–212
Le Carré, John, 181
Le Guin, Ursula, 232, 236
Lewontin, Richard, 134
Liboiron, Max, 56
Life on Earth (TV show), 34–35
"Like a Tumbleweed in Eden" (song, Robinson), 211–212
Linnaeus, Carl, 2, 3–4, 27, 81; binomial nomenclature and, 84–89; characterization of African women, 156; legacy of, in botany, 140–143; Linnaean labyrinths, 2–4; marriage of plants, 2, 9, 137, 140, 227; *Philosophia Botanica*, 85; plant reproductive biology and, 135–143; as promoter of his own work, 87–88; on race, 186; *Species Plantarum*, 3, 139; *Systema Naturae*, 3, 84–85; tree of life and, 157–158
Lipsitz, George, 237
livingness, 150
local ecologies, 46, 195
local knowledges, 21, 45–46, 117–118
Lorde, Audre, 5, 49, 75, 152, 230, 232, 233
Lose Your Mother (Hartman), 23

Lowe, Lisa, 163–164
Lowman, Meg, 8
Lucy (Kincaid), 33
Lugones, Maria, 155–156
Lyons, Kristina, 231

Machine in the Garden (Marx), 44
Mahābhārata, 29
Malthus, Thomas, 9, 22
Manilal, K. S., 107, 108, 112
Map to the Door of No Return, A (Brand), 153
Marder, Michael, 45
Margulis, Lynn, 101, 151
Markowitz, Sally, 159
marriage of plants. *See under* Linnaeus, Carl
Marx, Leo, 44
Mayr, Ernst, 80, 83, 138
McClintock, Anne, 100
McClintock, Barbara, 41, 101
McKittrick, Katherine, 127, 129–130, 150, 152, 231
Meeropool, Abel, 154
Mehta, Suketu, 183, 190
Meloy, Ellen, 131
Mendoza, Breny, 155
Menon, Minakshi, 122
Merchant, Carolyn, 45
Metaphor Project, 171
Methodology of the Oppressed (Sandoval), 229
methodus propria (Linnaeus), 86
Mignolo, Walter, 98
Mohanty, Chandra, 152
molecular biology, 90
monoculture of knowledge, 206
Morrison, Toni, 8–9, 135, 152, 191, 232
Mortimer-Sandilands, Catriona, 231
Most Wanted Man, A (Le Carré), 181
"Mother Nature's Melting Pot" (Raffles), 210
Muir, John, 44
Mukharji, Projit, 117–118
Muñoz, José Esteban, 152
Murphy, Michelle, 231
Myers, Natasha, 152, 231

naming, plant, 21–22, 79–80; colonialism and, 86–89; infrastructures of nomenclatures, classifications, and, 92–94; politics of, 209–211; postcolonial, 101–102; prehistory of, 81–84. *See also* nomenclature
Namjoshi, Suniti, 232
nationalism, 19, 53, 93, 185, 190, 213, 221, 223; Hindu, 57–58, 78, 117, 121, 224
native knowledge. *See* Indigenous knowledge
nativeness (concept), 193–196, 207; embracing contradictions in, 211–212; politics of naming and, 209–211
natural selection, 17, 35, 90, 149, 167–168
Nature (journal), 210
naturecultures, 6, 39–42, 48, 152, 154, 184, 202, 207–209
Nature's Body (Schiebinger), 135, 138
Nelson, Eli, 56
New York Times, 210
Nixon, Rob, 166
Noble, David, 236
nomenclature, 80, 139–140; coloniality of, 104, 120–122; folk taxonomies and prehistory of, 81–84; *Hortus Malabaricus*, 106–113; infrastructures of naming, classifications, and, 92–94; Linnaeus and legacy of binomial, 84–89; plant reproductive biology, 142–143. *See also* naming, plant
nuptaiae plantarum (Linnaeus), 2, 137

Ogden, Laura A., 198
O'Grady, Lorraine, 153
Olympia's Maid (O'Grady), 153
On the Origin of Species (Darwin), 34
orchids, 146–148, 152, 284n70

Osseo-Asare, Abena Dove, 96
overdevelopment, 192–193, 210, 228

Pandit, Vinayak, 110
Pangaea, ix, 183–184, 239–240
Panglossian Paradigm, 134
"parachute" science, 20, 95, 122, 249
Pauly, Philip, 207
"Pelvic Politics" (Markowitz), 159
Pérez-Bustos, Tania, 58, 238
"Perhaps the World Ends Here" (Harjo), 105
pharmaceuticals, plant-based, 113–118
Philosophia Botanica (Linnaeus), 85
phylogeny, 90–91, 92, 100–102
planet, 135, 177–178, 202, 204, 215, 218, 233, 239; globalization of the, 224; resilience of the, 236
planetarity, 231; queer, 174–178
planetary consciousness, 140, 238
planetary tapestry, 236
planetary time, 183–185, 240–242
plant reproductive biology, 133–134; botany and legacy of Linnaeus in, 140–143; do plants have sex, 148–150; gynodioecious plants, 142; how plants have sex, 143–148; livingness and, 150; in orchids, 146–148; reasons for plants having sex, 135–143; self-fertilization and, 148–149
plants: adaptation by, 35; colonialism controlling, 161–165; diasporas and, 201–202; fieldwork on, 40–41; hermaphroditic, 141–143, 145; history of pressed, 49–50; naming of (*see* naming, plant; nomenclature); nativeness of, 193–196, 207; new tree of life and, 165–167; pharmaceuticals using, 113–118; philosophy and love of, 45; renaming of, 21–22; Sanskrit names for, 98
Pollan, Michael, 22
Pollution Is Colonialism (Liboiron), 56
Polwhele, Richard, 141
polygamodioecious plants, 142
polygamomonoecious plants, 142
postcolonialism, 57–58, 120–121, 122–123, 202; naming and, 101–102
Prasad, Chandra Bhan, 120
Pratt, Mary Louise, 140
Prescod-Weinstein, Chanda, 231
pressed plants, 50. *See also* herbaria
Prosek, James, 87
purity, 22, 42, 52, 55, 102, 118, 121, 205, 228, 244; impurity vs., 111, 235; politics of, 78, 111, 165, 167, 222, 234–235; of local "authentic" cultures, 196

queer and trans communities, 28
queer biology, 168–169
queer botany, 10, 20, 22
queer ecology, 25, 151, 165, 167, 231
queering nature, 25
queer studies, 11–14, 48, 51, 165, 174
queer time, 13
Queer Vegennials, 170–173
Quijano, Aníbal, 54, 98, 155, 186

race: and coloniality of power, 186; eugenics, 11, 127, 160, 175, 233; feminist study of, 158–160; tied to gender, 100–101, 139; tree of life and, 157–158
racial capitalism, 11, 163–164
Raffles, Hugh, 210
Raj, Kapil, 109, 112
Raman, C. V., 29, 70
Ramanujan, Srinivasa, 29
Ramaswamy, E. V., 119
Ramayana, 29, 30
Raucous Spring, 61–62, 126, 173, 213
Ray, John, 27, 81
Reardon, Jenny, 101
recolonization, 57
"Re-imagining Reproduction" (Subramaniam and Bartlett), 144
Reinventing Hoodia (Foster), 96
rematriation, 20, 53, 72, 261n69

rememory, 135, 152, 192, 198, 227, 233; concept of, 8–9; usefulness of, 10, 232
Remesh, M., 112
renaming of plants, 21–22
Reo, Nicholas J., 198
reproductive biology: control over, 160–169; feminist study of, 158–160; racialized nature of, 167–168. *See also* plant reproductive biology
reproductive politics, 161
Rhee, Jeong-eun, 9
Rheede, Hendrik van, 107–110, 113, 120–121
rhizomatic networks, 201, 232
rhizomes, 171, 201, 232
Rhodes, Cecil John, 97
Rhodes Must Fall campaign, 97
Roberts, Dorothy, 167
Robinson, Chris, 211–212
Rodney, Walter, 195, 234
Roughgarden, Joan, 8, 150–151, 168
Rousseau, Jean Jacques, 3
Roy, Arundhati, 120, 227, 231
Roy, Sumana, 12

Sagan, Carl, 35–36
Sagan, Dorian, 151
Saleh-Hanna, Viviane, 8
Sandoval, Chela, 228, 229–230
Sanskrit names, 98, 112
Santiniketan, India, 218–224
Satia, Priya, 50, 238
Schiebinger, Londa, 135–136, 137, 138, 205
Schiwy, Freya, 155
science: abolitionist, 247–252; adisciplinary, 227–235; anti-Edenic, 243–246; coloniality of, 47, 51–54; culture of, 36–39; environmental and natural, 44–45; ethno-, 6; eugenic, 243; evolutionary, 89–91; gender and race in, 100–101; interdisciplinarity of, 6–7, 8, 47–48; local knowledges in, 21, 45–46; "parachute," 20; queer studies, disability studies, and, 11–14; sublime worlds of, 34–36; Traditional Ecological Knowledge (TEK), 7; universal, 71; women in, 65–73. *See also* botany; settler science
scientism, 251
scientization: of botany, 18, 137; of environmentalism, 175
Seasons (Meloy), 131
selection: natural, 17, 35, 90, 149, 167–168; sexual, 146, 151
self-fertilization, 148–149
sensing, 202–205
September 11, 2001, attacks, 184, 209
settler colonialism, 16, 284n71; history and endurance of, 15, 53, 156, 174, 247; impacts of, 175, 203, 213; vs. postcolonialism and decolonialism, 56, 57; reproduction and, 158; as rewriting history, 190, 191; in US imagination of nature, 44. *See also* nativeness (concept)
settler logics, 191, 193, 198
settler science, 56, 198, 243, 246
sex, plant. *See* plant reproductive biology
sexuality: feminist study of, 158–160; gynodioecious plants, 142; multiple possibilities in, 291n7; racialized nature of, 167–168; Victorian, 147–148
sexual selection, 146, 151
Shah, Amit, 120
Sharma, Kriti, 8
Sharpe, Christina, 9
Shaw, Philip, 43
Shotwell, Alexis, 77, 231
Shteir, Ann, 2–3
Siegesbeck, Johann Georg, 85, 141
Sierra Club, 97
Sikander, Shahzia, 229
silkworms, 124–125
Simpson, Audra, 54
Simpson, Leanne, 51–52, 150, 183, 196, 229, 230

Sims, Marion, 159–160
Singer, Emily, 102
Singh, Charu, 122
Sister Outsider (Lorde), 49, 75
sixth extinction, 95
Slime Mold Project, 172–173
Smith, Linda Tuhiwai, 73
Smith, Zadie, 199, 200
South Asia: kitchen space in, 105–106; spices and colonization of, 103–104. *See also* India
Species Plantarum (Linnaeus), 3, 139
spices, 103–104; as a "resource curse," 104; in the South Asian kitchen, 105–106, 127–128
Spivak, Gayatri, 50, 202, 231
Standing Rock protests, 166
Starr, Susan Leigh, 92
Star Trek (TV show), 34
state, repression by the, 54–55
Stepan, Nancy, 159
sterilization abuse, 160
"Strange Fruit" (song, Meeropool), 154
Stuessy, Tod, 83
subdioecious plants, 142
subgynoecious plants, 142
sublime, the, 26–27; defining, 42–44; otherworldliness and, 30–31; in science, 34–36
sublimity, 43
supercrip, 12
Systema Naturae (Linnaeus), 3, 84–85
systematics, 80; cladistics and, 91

Tagore, Rabindranath, 218–224; on nature, 221–224; universal humanism of, 220–221
TallBear, Kim, 49, 57, 197, 229, 231
Tang-Martinez, Zuleyma, 146
taxonomy, 80; folk, 81–84; imperial archive and, 94; Linnaean, 137, 139, 244; plant reproductive biology, 138
terraforming, 58–59
Their Eyes Were Watching God (Hurston), 133
Theophrastus, 81
This Land Is Our Land (Mehta), 183
Thoreau, Henry David, 44, 45
Time's Monster (Satia), 50
Tomes, Nancy, 185
"To the Diaspora" (Brooks), 212
Tournefort, Joseph Pitton de, 81
Trade Related Aspects of Intellectual Property Rights (TRIPS), 114
Traditional Ecological Knowledge (TEK), 7
Traditional Knowledge Database Library (TKDL), 113–118, 120–121
trans animals, 151
trans communities of color, 123
trans ecologies, 13
trans feminists, 8
trans movements, 166
trans nationalism, 122, 170
trans students, 65
trans studies, 13, 14, 123, 155
Traweek, Sharon, 39
tree of life, 90–91, 100–102; plant studies and the new, 165–167; theory behind, 157–158
trimonoecious plants, 143
trioecious plants, 143
Trivers, Robert, 146
"Trouble with Wilderness, The" (Cronon), 43
Tsing, Anna Lowenhaupt, 231
Tuck, Eve, 230–231, 238
tumbleweeds, 199–200, 204
turmeric, 103, 104, 113–114, 115, 127, 128
"types," plant, 96, 102, 108

Undrowned (Gumbs), 156
universal humanism, 220–221
universalism, 221; anti-, 244–245
universal knowledge, 248
universal science, 71
"Unsex'd Females, The" (Polwhele), 141

Valdez, Natali, 164
van der Weele, Cor, 195
Vazquez, Rolando, 55–56
Vedic knowledge, 111, 117
Veeramani, K., 119
Velocci, Beans, 167
Verghese, Abraham, 25
Victorian sexuality, 147–148
Voltaire, 134

wake work (Sharpe), 9
Wall Street Journal, 209
Walsh, Catherine, 155
Watanabe, Tsutomu, 101
Wayne, Marta, 146
Weinbaum, Alys E., 161
Weinreich, Max, 78
Weisinger, Marsha, 231
white empiricism, 231
White Teeth (Smith), 199
"Why Mammals Are Called Mammals" (Schiebinger), 205
wilderness, 6, 42–44, 55, 59, 129, 293n16; Adam and Eve banished to, 30, 59; David Attenborough on, 35; sounds of, 129
Willey, Angela, 13, 152, 231
Williams, G. C., 145
Williams, Patricia, 228–229
Wilson, E. O., 94, 95
Wilson, Elizabeth, 152, 168
Woese, Carl, 101
Wölfle Hazard, Cleo, 8, 231
Wollstonecraft, Mary, 141
women in science, 65–73
women of color, 105–106, 164; feminism and, 8, 244
Wordsworth, William, 44
World Bank, 193
World Council of Indigenous Peoples, 101
Wretched of the Earth, The (Fanon), 49
Wyer, Mary, 39, 47
Wynter, Sylvia, 10, 98–99, 152, 154–155, 186

xenophobia, 184, 185, 187, 192, 194, 208–209; politics of naming and, 209–211

Yang, Wayne, 238

FEMINIST TECHNOSCIENCES

Rebecca Herzig and Banu Subramaniam / Series Editors

Figuring the Population Bomb: Gender and Demography in the Mid-Twentieth Century, by Carole R. McCann

Risky Bodies and Techno-Intimacy: Reflections on Sexuality, Media, Science, Finance, by Geeta Patel

Reinventing Hoodia: Peoples, Plants, and Patents in South Africa, by Laura A. Foster

Queer Feminist Science Studies: A Reader, edited by Cyd Cipolla, Kristina Gupta, David A. Rubin, and Angela Willey

Gender before Birth: Sex Selection in a Transnational Context, by Rajani Bhatia

Molecular Feminisms: Biology, Becomings, and Life in the Lab, by Deboleena Roy

Holy Science: The Biopolitics of Hindu Nationalism, by Banu Subramaniam

Bad Dog: Pit Bull Politics and Mulitspecies Justice, by Harlan Weaver

Underflows: Queer Trans Ecologies and River Justice, by Cleo Wölfle Hazard

Hacking the Underground: Disability, Infrastructure, and London's Public Transport System, by Raquel Velho

Queer Data Studies, edited by Patrick Keilty

Botany of Empire: Plant Worlds and the Scientific Legacies of Colonialism, by Banu Subramaniam